Lecture Notes in Computer Science 12967

More information about this subseries at http://www.springer.com/series/7412

J. Alison Noble · Stephen Aylward ·
Alexander Grimwood · Zhe Min · Su-Lin Lee ·
Yipeng Hu (Eds.)

Simplifying Medical Ultrasound

Second International Workshop, ASMUS 2021
Held in Conjunction with MICCAI 2021
Strasbourg, France, September 27, 2021
Proceedings

Springer

Editors
J. Alison Noble 🄳
University of Oxford
Oxford, UK

Alexander Grimwood 🄳
University College London
London, UK

Su-Lin Lee 🄳
University College London
London, UK

Stephen Aylward 🄳
Kitware Inc.
Carrboro, NC, USA

Zhe Min 🄳
University College London
London, UK

Yipeng Hu 🄳
University College London
London, UK

ISSN 0302-9743   ISSN 1611-3349 (electronic)
Lecture Notes in Computer Science
ISBN 978-3-030-87582-4   ISBN 978-3-030-87583-1 (eBook)
https://doi.org/10.1007/978-3-030-87583-1

LNCS Sublibrary: SL6 – Image Processing, Computer Vision, Pattern Recognition, and Graphics

This Springer imprint is published by the registered company Springer Nature Switzerland AG
The registered company address is: Gewerbestrasse 11, 6330 Cham, Switzerland

Preface

In this exciting era for medical ultrasound, recent developments in deep learning (artificial intelligence) and medical robotics have started to show clinically measurable improvement in assisting ultrasound examinations, ultrasound-guided interventions, and surgery. ASMUS 2021: the 2nd International Workshop of Advances in Simplifying Medical UltraSound provided a forum for research topics around ultrasound image computing and computer-assisted interventions and robotic systems that utilize ultrasound imaging. ASMUS is the official workshop of the MICCAI Special Interest Group on Medical Ultrasound.

We were pleased to see the high quality of the submissions, evidenced by reviewer feedback and the fact that we had to reject a number of papers with merit due to the competitive nature of this workshop. The 22 accepted papers were selected based on their scientific contribution, via a double-blind process involving written reviews from at least two external reviewers in addition to a member of the Program Committee. Partly due to the focused interest and expertise, we received a set of exceptionally high-quality reviews and consistent recommendations.

The published work includes reports across a wide range of methodology, research, and clinical applications. The research topics cover advanced deep learning approaches for anatomy segmentation, lesion detection, ultrasound registration, ultrasound probe localization, robot-guided ultrasound, classification, image synthesis, quality assessment, and uncertainty estimation, with applications to point-of-care ultrasound systems and scenarios as well as computer-assisted interventions.

We would like to thank all reviewers, organizers, authors, and our keynote speakers, Muyinatu A. Lediju Bell, Ali Kamen, and Prerna Dogra, for sharing their research and expertise with our community, and we look forward to this workshop inspiring many exciting developments in the future in this important area.

September 2021

J. Alison Noble
Stephen Aylward
Alexander Grimwood
Zhe Min
Su-Lin Lee
Yipeng Hu

Organization

Organizing Chairs

J. Alison Noble   University of Oxford, UK
Stephen Aylward   Kitware, USA
Yipeng Hu   University College London, UK

Program Chairs

Alexander Grimwood   University College London, UK
Zhe Min   University College London, UK
Su-Lin Lee   University College London, UK

Demonstration Chair

Zachary Baum   University College London, UK

Organizing Committee

Ana Namburete   University of Oxford, UK
Andrew King   King's College London, UK
Bernhard Kainz   Imperial College London, UK
Dong Ni   Shenzhen University, China
Ekaterina Zilonova   KU Leuven, Belgium
Emad Boctor   Johns Hopkins University, USA
Parvin Mousavi   Queen's University, Canada
Purang Abolmaesumi   University of British Columbia, Canada
Thomas van den Heuvel Radboud University Medical Center, The Netherlands
Wolfgang Wein   ImFusion GmbH, Germany

Advisory Committee

Chris de Korte   Radboud University Nijmegen, The Netherlands
Gabor Fichtinger   Queen's University, Canada
Jan d'Hooge   KU Leuven, Belgium
Kawal Rhode   King's College London, UK
Nassir Navab   Technical University of Munich, Germany
Russ Taylor   Johns Hopkins University, USA

Program Committee

Stephen Aylward	Kitware, USA
Shuangyi Wang	Chinese Academy of Sciences, China
Bernhard Kainz	Imperial College London, UK
Francisco Vasconcelos	University College London, UK
Zhe Min	University College London, UK
Bradley Moore	Kitware, USA
Ilker Hacihaliloglu	Rutgers University, USA
Ana Namburete	University of Oxford, UK
Mohammad Alsharid	University of Oxford, UK
J. Alison Noble	University of Oxford, UK
Yuan Gao	University of Oxford, UK
Keshav Bimbraw	Worcester Polytechnic Institute, USA
Andrew Gilbert	GE Healthcare, Norway
Thomas van den Heuvel	Radboud University Medical Center, The Netherlands
Matthew Holden	Carleton University, Canada
Nina Montana-Brown	University College London, UK
Su-Lin Lee	University College London, UK
Yipeng Hu	University College London, UK
Parvin Mousavi	Queen's University, Canada
Emad Boctor	Johns Hopkins University, USA
Wietske Bastiaansen	Erasmus MC, The Netherlands
Joao Ramalhinho	University College London, UK
Ester Bonmati	University College London, UK
Zhen Yuan	King's College London, UK
Dong Ni	Shenzhen University, China
Jeremy Tan	Imperial College London, UK
Qianye Yang	University College London, UK
Shaheer Ullah Saeed	University College London, UK
Aaron Fenster	Robarts Research Institute, Canada
Lasse Lovstakken	Norwegian University of Science and Technology, Norway
Alexander Grimwood	University College London, UK
Ruoxi Gao	University College London, UK
Yipei Wang	University of Oxford, UK
Liansheng Wang	Xiamen University, China
Clare Teng	University of Oxford, UK
Zachary Baum	University College London, UK
Iani Gayo	University College London, UK
Ekaterina Zilonova	KU Leuven, Belgium
Harshita Sharma	University of Oxford, UK
Yunguan Fu	University College London, UK
Andrew King	King's College London, UK
Orcun Goksel	Uppsala University, Sweden
Wolfgang Wein	ImFusion GmbH, Germany
Jurica Sprem	GE Healthcare, Norway

Sponsors

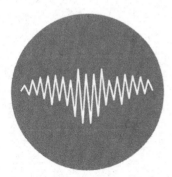

ThinkSono Ltd.

Ultromics Ltd.

Wellcome/EPSRC Centre for Interventional and Surgical Sciences (WEISS)

Nvidia Corporation

Contents

Registration, Guidance and Robotics

Classification and Image Synthesis

Quality Assessment and Quantitative Imaging

Segmentation and Detection

Automatic Ultrasound Vessel Segmentation with Deep Spatiotemporal Context Learning

Baichuan Jiang[1,2], Alvin Chen[1], Shyam Bharat[1], and Mingxin Zheng[1(✉)]

[1] Philips Research North America, 222 Jacobs Street, Cambridge, MA 02141, USA
`bcjiang@jhu.edu`, {`alvin.chen,shyam.bharat,mingxin.zheng`}`@philips.com`
[2] Department of Computer Science, Johns Hopkins University,
Baltimore, MD 21218, USA

Abstract. Accurate, real-time segmentation of vessel structures in ultrasound image sequences can aid in the measurement of lumen diameters and assessment of vascular diseases. This, however, remains a challenging task, particularly for extremely small vessels that are difficult to visualize. We propose to leverage the rich spatiotemporal context available in ultrasound to improve segmentation of small-scale lower-extremity arterial vasculature. We describe efficient deep learning methods that incorporate temporal, spatial, and feature-aware contextual embeddings at multiple resolution scales while jointly utilizing information from B-mode and Color Doppler signals. Evaluating on femoral and tibial artery scans performed on healthy subjects by an expert ultrasonographer, and comparing to consensus expert ground-truth annotations of inner lumen boundaries, we demonstrate real-time segmentation using the context-aware models and show that they significantly outperform comparable baseline approaches.

Keywords: Deep learning · Spatiotemporal attention · Vascular ultrasound

1 Introduction

Over 120 million people worldwide are affected by peripheral vascular disease, making it one of the leading causes of morbidity and mortality globally [1]. Vascular ultrasound (US) based on B-mode and Color Doppler imaging is widely used to evaluate luminal narrowing and flow in stenotic vessel segments. Accurate delineation of vessel wall boundaries is crucial for diagnosis and disease prognosis [2], but can be highly challenging due to complex vascular anatomy, visual ambiguity, acoustic imaging artefacts, probe motion, and very small target

---

B. Jiang and A. Chen—Contributed equally.
Work done during internship at Philips Research.

J. A. Noble et al. (Eds.): ASMUS 2021, LNCS 12967, pp. 3–13, 2021.
https://doi.org/10.1007/978-3-030-87583-1_1

structures hidden within the US image frame [3]. Automated, real-time segmentation of relevant vessels can aid in clinical assessment while providing a means to improve vascular US imaging workflows and reduce operator dependency.

Prior work on US vessel segmentation have utilized shape and motion models [4–7] typically requiring initialization with seed points in the first frame. To provide additional context, the inclusion of flow information alongside the B-mode image has been proposed [8–11]. In recent years, deep learning has received increased focus [2,12], with the majority of work on vessel segmentation utilizing UNet/VNet-like models operating on individual frames [13,14]. Incorporating the time dimension, recurrent mechanisms have been combined with convolutional networks as a way to encode temporal memory from image sequences [15–20]. However, to date, most studies have focused on segmentation of large carotid [13,14] or coronary [20] arteries using UNet/VNet, and accuracy on small extremity vasculature has not been systematically investigated.

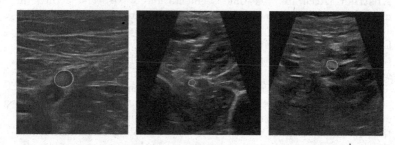

Fig. 1. B-mode and Color Doppler US images of lower-extremity arterial vasculature. Expert ground-truth annotations of inner lumen vessel wall boundaries shown in green. *Left*: femoral artery (∼5 mm diameter), *Middle*: anterior tibial artery (∼2 mm diameter), *Right*: posterior tibial artery (∼3 mm diameter). (Color figure online)

Our Contributions. We demonstrate that a spatiotemporally-aware deep learning model is capable of automatic, real-time segmentation of inner lumen vessel boundaries from challenging freehand US sequences (Fig. 1). Improved performance in difficult anatomy, namely small-scale lower-extremity peripheral arterial vasculature, is made possible by leveraging the rich spatiotemporal information available in US and utilizing dual-input B-mode and color flow signals. Our approach aims to simulate the contextual inferencing processes of experienced ultrasonographers, who are trained to recognize temporal signatures in both modalities and attend to small structures of interest while ignoring background.

Specifically, we propose a fully convolutional encoder-decoder network that feeds B-mode and Color Doppler inputs through a series of spatial, temporal, and channel-wise contextual units embedded within each resolution layer. By preserving multi-resolution features as they pass through each unit, the network is able to propagate learned representations temporally across all scales.

Additional Contributions of the Paper:

1. We systematically study the impact of the multi-scale spatial, temporal, and channel-wise embeddings for segmentation of femoral (4–6 mm diameter) and tibial (2–3 mm diameter) arteries in a series of network ablation experiments.
2. We investigate the utility of exploiting combined B-mode and Color Doppler information compared to B-mode alone, and we evaluate the benefits of domain-specific augmentation on the US small-structure segmentation task.
3. We compare segmentation results against consensus ground-truth annotations from multiple clinical experts, and we demonstrate significant improvements in accuracy using the proposed methods compared to baseline models.

2   Methodology

Model Overview. The proposed VESsel NETwork with Spatial, Channel and Temporal context (VesNetSCT+) is illustrated in Fig. 2. The network architecture and implementation details are described below.

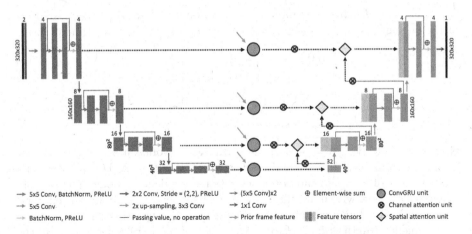

Fig. 2. Network architecture. The model feeds B-mode and Color Doppler inputs through a multi-scale series of spatial, temporal, and feature-wise contextual units embedded in each resolution layer of a fully convolutional encoder-decoder backbone. The design enables efficient learning of multi-modal spatiotemporal context information for challenging small-structure segmentation tasks.

Encoder-Decoder Backbone. Our model uses a UNet/VNet-like [21,22] backbone with a two-channel input to support B-mode and Color Doppler frames. The proposed spatial, temporal, and feature/channel-wise contextual units are sequentially embedded along the residual skip connections [23] in each resolution layer. This design allows the model to aggregate learned representations across multiple spatial scales and time points. We additionally posit that, unlike methods which explicitly modify the backbone network [15–18], embedding into the skip connections minimizes disruption of gradients along the main path.

Implementation Details: B-mode and Color channels are normalized between (0, 1) and (−1, 1), respectively. We apply batch normalization and PReLU, and we use resize-up-convolutions [24] in the decoder to minimize checkerboard artifacts. We also reduce the total model size from 30M (original UNet) to 0.3M by opting for a four-layer network backbone capable of real-time inference on clinical US machines. Networks are trained via RMSProp with initial learning rate of 0.0001.

Multi-scale Temporal Gating. To introduce temporal context, a hierarchical series of convolutional gated recurrent units (ConvGRU) [25,26] are embedded between the contracting and expanding paths (round orange blocks in Fig. 2). By replacing dot product operations found in standard GRU with convolutional operations, the learned temporal representations have the inherent spatial connectivity of convolutional networks and are more parameter-efficient [26].

Implementation Details: The inputs into the ConvGRU are the feature maps x_t for the current frame t and the hidden states h_{t-1} from the prior time point:

$$z_t = \sigma(W_{hz} * h_{t-1} + W_{xz} * x_t + b_z) \tag{1}$$

$$r_t = \sigma(W_{hr} * h_{t-1} + W_{xr} * x_t + b_r) \tag{2}$$

$$\hat{h}_t = \tanh(W_h * (r_t \odot h_{t-1}) + W_x * x_t + b) \tag{3}$$

$$h_t = (1 - z_t) \odot h_{t-1} + z \odot \hat{h}_t \tag{4}$$

where $W_{hz}, W_{xz}, W_{hr}, W_{xr}, b_z, b_r, b$ are learnable convolution kernels and biases. z_t and r_t are internal update and reset gates. The output h_t is computed as a weighted sum of h_{t-1} and candidate activation $\hat{h}_t$. σ denotes sigmoid activation, $\odot$ denotes element-wise multiplication, and $*$ denotes convolution. All temporal models are trained via truncated backpropagation-through-time (TBTT) with a fixed sequence length of 50 frames. A time window of 1–4 frames is used in the feedforward step and accumulated to compute back-propagation gradients [27].

Multi-scale Spatial Attention. We add soft self-attention gates to provide higher-resolution layers with spatial support from coarse layers. This idea has been shown to improve accuracy on small-structure segmentation tasks [28] and is fully differentiable, unlike hard attention mechanisms based on iterative region proposal and cropping [29]. The spatial self-attention units are introduced in a multi-scale manner along the skip connections, immediately following the temporal and channel operators. The arrangement provides the self-attention gates with access to global spatiotemporal context when learning to suppress irrelevant background and attend to regions of interest.

Implementation Details: Within this unit, input feature maps x, and input gating feature maps g from the previous coarse resolution level, are used to derive attention multiplier maps $\alpha \colon [0,1] \to \mathbb{R}^2$. The self-attention maps are multiplied element-wise into each channel of x to produce the output maps $\hat{x}$:

$$\alpha = \sigma(\psi(\delta(W_x * x + W_g * g + b_g)) + b_\psi) \tag{5}$$

$$\hat{x}_c = x_c \odot \alpha \tag{6}$$

where σ and δ denote sigmoid and ReLU activations, respectively. W_x, W_g, ψ are weight parameters for the channel-wise $1 \times 1 \times 1$ convolutions, b_ψ, b_g are the bias terms in the gating unit, and $\odot$ denotes element-wise multiplication between the multiplier and input feature maps at channel c.

Multi-scale Feature-Wise Channel Attention. Finally, feature/channel self-attention is incorporated in the skip connections and along the decoder path. Specifically, we employ convolutional block attention modules [30] immediately before each spatial attention gate (round green blocks in Fig. 2). Particularly for tasks involving multiple input modalities, self-attention along the channel dimension offers a mechanism to explicitly model interdependencies between modalities (in our case, the encoded B-mode and Color Doppler input signals).

Implementation Details: Given input feature maps x^c with c channels, average- and max-pooling are performed along the channel dimension to produce descriptors x^c_{avg}, $x^c_{max} \in \mathbb{R}^{1 \times 1 \times C}$. These are passed through a shared multi-layer perceptron (MLP) g and element-wise summed. The result, a vector $m_c \in \mathbb{R}^{1 \times 1 \times C}$, is channel-wise multiplied ($\otimes$) with the input x^c:

$$\hat{x}^c = m_c \otimes x^c \tag{7}$$

$$= \{\sigma(g(x^c_{avg}) + g(x^c_{max}))\} \otimes x^c \tag{8}$$

$$= \{\sigma(W_1(\delta(W_0(x^c_{avg}))) + W_1(\delta(W_0(x^c_{max}))))\} \otimes x^c \tag{9}$$

where σ and δ denote sigmoid and ReLU, and W_0, W_1 are learnable MLP weights.

Domain-Specific Augmentation. Extensive data augmentation was applied in training to improve generalizability and robustness to real-world freehand US imaging conditions. Augmentations were defined on training sequences spanning 50 sequential frames, and included: (1) *Spatial augmentation* based on random translation, rotation, scaling, cropping, and horizontal flipping; (2) *Gain/contrast augmentation* based on random adjustment of histogram and time gain compensation curves separately to the B-mode and Color channels; (3) *Color Doppler augmentation*, where channel dropout is applied with fixed probability on the Color inputs to simulate poor Doppler signal due to impaired blood flow; and (4) *Temporal augmentation* by varying the start and end frames, interval between frames, and order of frames (forward or reverse) for each set of 50-frame inputs.

Vascular Ultrasound Data Acquisition. Freehand lower-extremity arterial US exams were performed by an expert vascular sonographer on left and right legs of 7 healthy subjects. Scans were acquired in transverse orientation following standard imaging workflows for diagnostic Duplex ultrasonography [31]. The scans were performed under simultaneous B-mode and Color Doppler modes and spanned the length of the leg, from ankle to groin. In total, 22 exam sequences were acquired, including 13 along the femoral arteries and 9 along the anterior/posterior tibial arteries. All data were collected with a Philips Epiq 7 system and 12 MHz linear transducer (pixel spacing ~0.1 mm, frame rate ~15 Hz).

Clinical Expert Ground-Truth Annotation. Expert ground-truth annotations of inner lumen boundaries (Fig. 1) were provided by two experienced vascular sonographers. Consensus ground-truth masks were computed via shape-based averaging of the two sets of individual segmentations [32]. A total of 30,839 consensus-annotated frames were obtained from the 22 femoral and tibial artery exams in this manner. The data were divided according to subject to allow for independent training and testing, and to perform leave-one-out cross-validation.

3   Results

Validation and Network Ablation Studies

Temporal, Spatial, and Channel-Wise Context: On both femoral and tibial hold-out test data, we saw significant improvements in Dice scores (Table 1) with the addition of temporal, spatial, and feature/channel-wise contextual embeddings alongside bimodal B-mode+Color inputs. For the femoral dataset, the best-performing context-aware model (VesNetSCT++, 0.927 ± 0.041 Dice) demonstrated an improvement of 18 to 20% in comparison to two baseline UNet models of varying sizes (Baseline, 0.775 ± 0.282 Dice; Baseline-L, 0.788 ± 0.301 Dice). Meanwhile, in the tibial arteries, which represented <0.1% of pixels in each frame, the baseline models failed entirely (Baseline, 0.133 ± 0.227 Dice; Baseline-L, 0.164 ± 0.254 Dice) compared to the context-aware model (VesNetSCT++, 0.679 ± 0.195 Dice).

Multi-scale Embeddings: The introduction of temporal/spatial/channel embeddings in a multi-scale manner (VesNetT+) resulted in improved performance compared to equivalent models with the embeddings applied only to the innermost layer of the encoder-decoder backbone (VesNetT), as proposed in [15–18].

Temporal Window: We experimented with the time window for accumulating feed-forward and back-propagation updates when training temporal models using TBTT [27]. Improvements were seen by increasing the window size (VesNetSCT++, time window = 4), at the cost of longer training times.

Table 1. Summary of segmentation performance on femoral and tibial test sets.

Model name*	# params	Input channels	Spatial/channel attention	Temporal gating	Time window	Dice score (mean ± std)
Femoral arteries (4–6 mm diameter)						
Baseline	103k	Bmode	-	-	-	0.775 ± 0.282
Baseline-L	310k	Bmode	-	-	-	0.788 ± 0.301
VesNet	103k	Bmode+Color	-	-	-	0.870 ± 0.150
VesNetS	105k	Bmode+Color	S	-	-	0.881 ± 0.175
VesNetSC	106k	Bmode+Color	S+C	-	-	0.903 ± 0.101
VesNet-L	313k	Bmode+Color	-	-	-	0.887 ± 0.180
VesNetT	259k	Bmode+Color	-	Single	1	0.908 ± 0.086
VesNetT+	307k	Bmode+Color	-	Multi-scale	1	0.914 ± 0.065
VesNetST+	309k	Bmode+Color	S	Multi-scale	1	0.919 ± 0.069
VesNetSCT+	310k	Bmode+Color	S+C	Multi-scale	1	0.925 ± 0.051
VesNetSCT++	310k	Bmode+Color	S+C	Multi-scale	4	**0.927 ± 0.041**
Tibial arteries (2–3 mm diameter)						
Baseline	103k	B-mode	-	-	-	0.133 ± 0.227
Baseline-L	310k	B-mode	-	-	-	0.164 ± 0.254
VesNet	103k	Bmode+Color	-	-	-	0.527 ± 0.336
VesNetS	105k	Bmode+Color	S	-	-	0.564 ± 0.317
VesNetSC	106k	Bmode+Color	S+C	-	-	0.570 ± 0.282
VesNet-L	313k	Bmode+Color	-	-	-	0.534 ± 0.328
VesNetT	259k	Bmode+Color	-	Single	1	0.564 ± 0.283
VesNetT+	307k	Bmode+Color	-	Multi-scale	1	0.655 ± 0.211
VesNetST+	309k	Bmode+Color	S	Multi-scale	1	0.664 ± 0.246
VesNetSCT+	310k	Bmode+Color	S+C	Multi-scale	1	0.671 ± 0.240
VesNetSCT++	310k	Bmode+Color	S+C	Multi-scale	4	**0.679 ± 0.195**

* *Nomenclature*. Baseline: UNet; VesNet: bimodal-input UNet; "-L": larger network with more channels per layer; "S": spatial attention; "C": channel attention; "T": temporal gating; "+": multi-scale embeddings; "++": expanded TBTT window.

Domain-Specific Augmentation: Table 2 compares models trained with and without channel dropout on the Color Doppler input. Table 3 compares results with and without temporal augmentation. Overall, we saw that the augmentations resulted in improved test accuracy, which suggests the model's robustness to the poor Doppler signal quality and varying freehand scanning motions.

Model Size and Baseline Comparisons: Comparing baseline UNet models modified to match the number of parameters as our final proposed models (Baseline-L, 0.3M parameters), a three-fold increase in parameters gave no appreciable improvement in performance (Table 1). This suggests that the improved accuracy of the context-aware models was not due simply to larger network size.

Table 2. Impact of color augmentation. See Table 1 nomenclature.

Model name	Color doppler dropout	Dice score
Femoral arteries (4–6 mm diameter)		
VesNetSC	0.0	0.880 ± 0.145
VesNetSC	0.4	0.903 ± 0.101
VesNetSCT+	0.0	0.911 ± 0.087
VesNetSCT+	0.4	**0.925 ± 0.051**
Tibial arteries (2–3 mm diameter)		
VesNetSC	0.0	0.579 ± 0.312
VesNetSC	0.4	0.570 ±0.282
VesNetSCT+	0.0	**0.676 ± 0.293**
VesNetSCT+	0.4	0.671 ± 0.240

Table 3. Impact of temporal augmentation. See Table 1 nomenclature.

Model name	Temporal augmentation	Dice score
Femoral arteries (4–6 mm)		
VesNetSCT+	No	0.917 ± 0.117
VesNetSCT+	Yes	**0.925 ± 0.051**
Tibial arteries (2–3 mm)		
VesNetSCT+	No	0.660 ± 0.245
VesNetSCT+	Yes	**0.671 ± 0.240**

Table 4. Cross-validation results with context-aware models (VesNetSCT+).

Data splits:	1	2	3	4	5	6	7	Mean ± Stdev
Femoral arteries	**0.925**	0.853	0.828	0.940	0.935	0.909	0.802	**0.885 ± 0.162**
Tibial arteries	**0.671**	0.782	0.565	0.511	0.646	-	-	**0.635 ± 0.215**

Cross Validation: To assess generalization, leave-one-out cross validation was carried out on both datasets using the contextually-aware models (VesNetSCT+). In each split, sequences from one subject were held out for testing (Table 4).

Inference Speed: VesNetSCT++ achieved inference speeds of 149.4 ± 4.6 ms (6.7 Hz) on a mobile CPU processor (Intel Core i7 2.6 GHz) and 8.9 ± 0.6 ms (112 Hz) on a mobile GPU (Nvidia RTX 2080). These speeds were significantly faster than those of the Baseline-L UNet model and show that the proposed methods are amenable to real-time processing on hardware used in clinical US machines.

Visualization of Results. Figure 3 shows examples of the effects of spatiotemporal context on hold-out sequences. In both cases, Dice scores fluctuate less when the contextual mechanisms are introduced. The addition of temporal gating (left panel, comparing VesNet and VesNetT+ models) allows the model to correctly learn pulsatile signatures and ignore confounding Doppler signals from nearby veins. With spatial attention gating (right panel, comparing VesNetT+ and VesNetST+), extremely small tibial arteries are more reliably localized throughout.

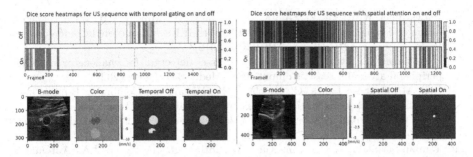

Fig. 3. Visualization of segmentations on hold-out test sequences. *Left*: Temporally-aware model (VesNetT+) outperforms an equivalent model operating in individual frames (VesNet). *Right*: Inclusion of spatial self-attention (VesNetST+) outperforms the same model without spatial attention (VesNetT+).

4   Conclusion

This work presented an efficient deep learning architecture that incorporates multiple strategies for embedding spatiotemporal context to improve segmentation of challenging 2D US image sequences. We applied the methods to small-scale lower extremity arteries from freehand B-mode and Color Doppler scans, and showed strong improvement over baseline models without the added contextual awareness. Future work will investigate the generalizability of these methods on other anatomies where flow and spatiotemporal information are available, and where automatic quantification of vascular measurements is of clinical benefit.

Acknowledgment. The authors thank Elizabeth Brunelle, Barbara Bannister, and Jochen Kruecker for assistance in data acquisition, annotation, and review.

References

1. Bauersachs, R., Zeymer, U., Brière, J.B., Marre, C., Bowrin, K., Huelsebeck, M.: Burden of coronary artery disease and peripheral artery disease: a literature review. In: Cardiovascular Therapeutics (2019)
2. Moccia, S., De Momi, E., El Hadji, S., Mattos, L.S.: Blood vessel segmentation algorithms – review of methods, datasets and evaluation metrics. Comput. Methods Programs Biomed. **158**, 71–91 (2018)
3. Liu, S., et al.: Deep learning in medical ultrasound analysis: a review. Engineering **5**, 261–275 (2019)
4. Guerrero, J., Salcudean, S.E., McEwen, J.A., Masri, B.A., Nicolaou, S.: Real-time vessel segmentation and tracking for ultrasound imaging applications. IEEE Trans. Med. Imag. **26**(8), 1079–1090 (2007)
5. Ma, L., Kiyomatsu, H., Nakagawa, K., Wang, J., Kobayashi, E., Sakuma, I.: Accurate vessel segmentation in ultrasound images using a local-phase-based snake. Biomed. Sig. Process. Control **43**, 236–243 (2018)

6. Patwardhan, K.A., Yu, Y., Gupta, S., Dentinger, A., Mills, D.: 4D vessel segmentation and tracking in ultrasound. In:2012 19th IEEE International Conference on Image Processing (ICIP), pp. 2317–2320 (2012)
7. Mistelbauer, G., et al.: Semi-automatic vessel detection for challenging cases of peripheral arterial disease. Comput. Biol. Med. **133**, 104344 (2021)
8. Keil, M., Oyarzun Laura, C., Drechsler, K., Wesarg, S.: Combining B-mode and color flow vessel segmentation for registration of hepatic CT and ultrasound volumes. In: Eurographics Workshop on Visual Computing for Biology and Medicine, pp. 57–64 (2012)
9. Tamimi-Sarnikowski, P., Brink-Kjær, A., Moshavegh, R., Jensen, J.A.: Automatic segmentation of vessels in in-vivo ultrasound scans. In: Proceedings of SPIE, p. 10137 (2017)
10. Moshavegh, R., Martins, B., Hansen, K.L., Bechsgaard, T., Bachmann Nielsen, M., Jensen, J.A.: Hybrid segmentation of vessels and automated flow measures in in-vivo ultrasound imaging. In: Proceedings of 2016 IEEE International Ultrasonics Symposium (IUS). IEEE (2016)
11. Akkus, Z., et al.: Fully automated carotid plaque segmentation in combined contrast-enhanced and B-mode ultrasound. Ultrasound Med. Biol. **41**(2), 517–531 (2015)
12. Smistad, E., Løvstakken, L.: Vessel detection in ultrasound images using deep convolutional neural networks. In: Carneiro, G., et al. (eds.) LABELS/DLMIA -2016. LNCS, vol. 10008, pp. 30–38. Springer, Cham (2016). https://doi.org/10.1007/978-3-319-46976-8_4
13. Zhou, R., et al.: Deep learning-based measurement of total plaque area in B-mode ultrasound images. IEEE J. Biomed. Health Inform. (2021). https://doi.org/10.1109/JBHI.2021.3060163
14. Zhou, R., Fenster, A., Xia, Y., Spence, J.D., Ding, M.: Deep learning-based carotid media-adventitia and lumen-intima boundary segmentation from three-dimensional ultrasound images. Med. Phys. **46**(7), 3180–3193 (2019)
15. Gao, Y., Phillips, J., Zheng, Y., Min, R., Fletcher, P., Gerig, G.: Fully convolutional structured LSTM networks for joint 4D medical image segmentation. In: Proceedings ISBI 2019, pp. 1104–1108. IEEE (2019)
16. Milletari, F., Rieke, N., Baust, M., Esposito, M., Navab, N.: CFCM: segmentation via coarse to fine context memory. In: Frangi, A.F., Schnabel, J.A., Davatzikos, C., Alberola-López, C., Fichtinger, G. (eds.) MICCAI 2018. LNCS, vol. 11073, pp. 667–674. Springer, Cham (2018). https://doi.org/10.1007/978-3-030-00937-3_76
17. Arbelle, A., Raviv, T.R.: Microscopy cell segmentation via convolutional LSTM Networks. In: Proceedings ISBI (2019)
18. Webb, J.M., Meixner, D.D., Adusei, S.A., Polley, E.C., Fatemi, M., Alizad, A.: Automatic deep learning semantic segmentation of ultrasound thyroid cineclips using recurrent fully convolutional networks. IEEE Access **9**, 5119–5127 (2020)
19. Gonzalez Duque, V., Al Chanti, D., Crouzier, M., Nordez, A., Lacourpaille, L., Mateus, D.: Spatio-temporal consistency and negative label transfer for 3D freehand US segmentation. In: Martel, A.L., et al. (eds.) MICCAI 2020. LNCS, vol. 12261, pp. 710–720. Springer, Cham (2020). https://doi.org/10.1007/978-3-030-59710-8_69
20. Mirunalini, P., Aravindan, C., Thamizh Nambi, A., Poorvaja, S., Pooja Priya, V.: Segmentation of coronary arteries from CTA axial slices using deep learning techniques. In: IEEE Region 10 International Conference (TENCON), pp. 2074–2080 (2019)

21. Ronneberger, O., Fischer, P., Brox, T.: U-Net: convolutional networks for biomedical image segmentation. In: Navab, N., Hornegger, J., Wells, W.M., Frangi, A.F. (eds.) MICCAI 2015. LNCS, vol. 9351, pp. 234–241. Springer, Cham (2015). https://doi.org/10.1007/978-3-319-24574-4_28
22. Milletari, F., Navab, N., Ahmadi, S.: V-Net: fully convolutional neural networks for volumetric medical image segmentation. In: 4th International Conference on 3D Vision (2016)
23. He, K., Zhang, X., Ren, S., Sun, J.: Deep residual learning for image recognition. In: Proceedings of the IEEE Conference on Computer Vision and Pattern Recognition (CVPR), pp. 770–778 (2016)
24. Odena, A., Dumoulin, V., Olah, C.: Deconvolution and checkerboard artifacts (2016). http://distill.pub/2016/deconv-checkerboard/
25. Siam, M., Valipour, S., Jagersand, M., Ray, N.: Convolutional gated recurrent networks for video segmentation. In: IEEE International Conference on Image Processing (ICIP), pp. 3090–3094 (2017)
26. Ballas, N., Yao, L., Pal, C., Courville, A.: Delving deeper into convolutional networks for learning video representations. arXiv:1511.06432 (2015)
27. Williams, R.J., Zipser, D.: Gradient-based learning algorithms for recurrent networks and their computational complexity. In: Chauvin, Y., Rumelhard, D.E. (eds.) Backpropagation: Theory, Architectures, and Applications (1995)
28. Oktay, O., et al.: Attention U-Net: Learning where to look for the pancreas. arXiv:1804.03999 (2018)
29. Mnih, V., Heess, N., Graves, A., Kavukcuoglu, K.: Recurrent models of visual attention. arXiv:1406.6247 (2014)
30. Woo, S., Park, J., Lee, J.Y., Kweon, I.S.: CBAM: convolutional block attention module. In: Proceedings of the European Conference on Computer Vision (ECCV), pp. 3–19 (2018)
31. Strandness, D.E., Jr.: Duplex Scanning in Vascular Disorders, 3rd edn. Lippincott Williams & Wilkins, Philadelphia, Pennsylvania (2001)
32. Rohlfing, T., Maurer, C.R.: Shape-based averaging for combination of multiple segmentations. In: Duncan, J.S., Gerig, G. (eds.) MICCAI 2005. LNCS, vol. 3750, pp. 838–845. Springer, Heidelberg (2005). https://doi.org/10.1007/11566489_103

Multimodal Continual Learning with Sonographer Eye-Tracking in Fetal Ultrasound

Arijit Patra[(✉)], Yifan Cai, Pierre Chatelain, Harshita Sharma, Lior Drukker, Aris T. Papageorghiou, and J. Alison Noble

University of Oxford, Oxford, Oxfordshire OX3 7DQ, UK
arijit.patra@eng.ox.ac.uk

Abstract. Deep networks have been shown to achieve impressive accuracy for some medical image analysis tasks where large datasets and annotations are available. However, tasks involving learning over new sets of classes arriving over extended time is a different and difficult challenge due to the tendency of reduction in performance over old classes while adapting to new ones. Controlling such a 'forgetting' is vital for deployed algorithms to evolve with new arrivals of data incrementally. Usually, incremental learning approaches rely on expert knowledge in the form of manual annotations or active feedback. In this paper, we explore the role that other forms of expert knowledge might play in making deep networks in medical image analysis immune to forgetting over extended time. We introduce a novel framework for mitigation of this forgetting effect in deep networks considering the case of combining ultrasound video with point-of-gaze tracked for expert sonographers during model training. This is used along with a novel weighted distillation strategy to reduce the propagation of effects due to class imbalance.

Keywords: Incremental learning · Eye tracking · Fetal ultrasound

1 Introduction

Deep networks often need large quantities of labeled data [16,21,22]. In medical imaging, large datasets may not always be available but collected over time [19,26]. Retention of past data over extended time is often difficult in medical imaging compared to natural images and similar datasets due to privacy concerns, statutory limitations on storage duration and memory constraints particular to clinical situations in different countries. Evolving diseases or diagnostic regulations may require models to adapt to new data classes arriving over time. This requires models to learn incrementally without declining in performance on prior tasks trained for. Deep networks in medical imaging have often tried to adapt to new tasks over time using transfer learning [2]. Recent work has shown that transfer learning, despite leveraging past learning, doesn't allow an effective balance between past learnt representations and current task knowledge

© Springer Nature Switzerland AG 2021
J. A. Noble et al. (Eds.): ASMUS 2021, LNCS 12967, pp. 14–24, 2021.
https://doi.org/10.1007/978-3-030-87583-1_2

owing to catastrophic forgetting [7], wherein neural network based models are seen to show a decreased performance on initially trained tasks when retrained on new tasks without access to initial task data. This requires regularization of the learning on current tasks using prior task knowledge. This has recently been introduced as the continual learning paradigm in medical image analysis. Human sonographers acquire knowledge over time without losing performance on previously learnt modules. Can sonographers' insights be used to improve knowledge transfer across sonography plane-finding tasks? This is explored with a novel multimodal incremental learning approach using class-weighted distillation. The question of multimodal information preservation in incremental adaptation remains unexplored in machine learning and medical imaging literature.

Recent work has attempted to reduce forgetting with replay on stored examples [14,25], expanding parameters [31], generative models [10,24] and weight regularization [13]. Besides model compression, knowledge distillation [8] was used in continual learning as the distillation loss is suitable for using a snapshot of learnt knowledge in a model at a particular learning step towards regularizing future learning. Recent methods using distillation for continual learning include Learning without Forgetting (LwF) [14], iCaRL [30] which incrementally performs representation learning, progressive distillation and retrospection (PDR) [9] and Learning without Memorizing (LwM) [4] where distillation is used with class activation. In medical imaging, real-world cases of data arriving over time has prompted research on continual pipelines such as MRI segmentation with pixel-level regularization [20], hierarchical learning in echocardiography [27], progressive modelling of Alzheimer's [32], consolidated distillation [11], distillation and ensembling [15] and privacy preserved learning [29]. Compared to these, our methods do not require retention of exemplars for past classes and implement a performance driven weighting for distilled representations. Usage of expert knowledge or other forms of multimodal input for incremental learning remains unexplored in medical imaging to our knowledge. Dedicated metrics for incremental learning [5] were proposed but [5] like most prior work, use accuracy-derived metrics for assessment. Kim et al. [11] compute AUROC scores to assess overall performances after all learning sessions but do not define forgetting by effects on 1-vs-all AUC in multiclass incremental settings. Sonographer eye-tracking was used for biometry plane localization [1], representation learning [6] and model compression [23]. Different from [23], we learn representations for incrementally learning new classes with the same trained model.

Contributions. We propose a framework that a) demonstrates the first usage of multimodal data of ultrasound frames along with expert gaze in incremental learning b) a novel weighted distillation strategy to reduce the impact of task-specific class imbalance over incremental learning c) proposes metrics for assessment of incremental learning using both accuracy and AUC measures. We achieve superior incremental learning performances in terms of mitigation of forgetting without storing any of the past data. The AUC is computed as area under precision-recall (PR) curves per class in a 1-vs-all setting. An alteration in the number of false positives, as possible due to catastrophic forgetting, may still cause a small change to false positive rate (used in ROC estimation) [3]. In PR curves, precision is obtained by comparing false positives to true positives, capturing effects of large numbers of false positives on incremental performance.

2   Methods

Incremental Learning. Without losing generality, let the tth stage in an M-stage incremental learning problem be a K-class classification task with a class set $\mathbf{X_t} = \{X_{t,i}\}_{i=1}^{K_t}, t \in [\![1, M]\!]$, where each X represents one class and training samples for stage t are drawn from the set: $x_{t,\text{train}} \in \mathbf{X_t}$. The objective is to recursively enable the classifier trained in stage $t - 1$, after completing the tth stage incremental learning pipeline, perform inference on examples $x_{\text{test}} \in \bigcup_{j=1}^{t}\mathbf{X}_j$ without declining performance. This is non-trivial if training data from previous stages are unavailable, making it impossible to jointly retrain the classifier on the entire dataset. The challenge is to ensure specific interventions are adopted to reduce catastrophic forgetting [7]. Here, $M = 2$ and $K_1 = K_2 = 3$. We consider a two-stage multitask incremental learning problem with Stage 1 aiming to learn a classifier for fetal biometry planes (head, femur, abdomen), while Stage 2 aims to expand the Stage 1 classifier for echocardiogrpahy tasks, *i.e.* identification of three fetal cardiac standard planes: frames of four chamber (4CH), three vessel (3VV) and left ventricular outflow tract (LVOT). We show that the final classifier doesn't suffer from forgetting for Stage 1 tasks without having to retrain using Stage 1 data. The output of the Stage 1 classifier $f_1(\cdot)$ can take the form $p = \text{softmax}(\mathbf{z}) \in \mathbb{R}^{K_1}$, where $\mathbf{z} = f_1(x_{1,train}) \in \mathbb{R}^{K_1}$ is the raw output of the last layer, or *logits*. The logits output is retained as old knowledge-based priors (similar to distillation [8]), and used in the 2nd-stage training as regularization for a new task of classifying fetal echocardiography standard planes.

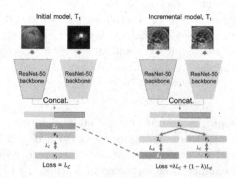

Fig. 1. The initially trained model (on Stage 1 tasks) is later trained for an incremental task at Stage t (here, t = 2), with cross-distillation using logits stored from initial stages.

Gaze Acquisition. For including multimodal inputs in incremental learning, we consider a paired input tuple *(I, G)* of images and corresponding gaze maps, with their associated labels for the initial training. To mimic situations where a gaze map *G* will not be available during future task training, subsequent stages do not assume presence of the gaze map and can function with the gaze map substituted with another modality or a redundant copy of the image. Similar to the protocol in [23], the expert visual attention is captured by a gaze map *G* per image *I*. The point-of-gaze of the expert is tracked when looking at *I*. The initial task session involves a classification task using a tuple of *I* and *G* as input. In subsequent incremental tasks, we allow for situations where models may have access to only the image frames of the new classes in line with the difficulty of acquiring gaze maps in deployment environments. Initial inclusion of gaze maps in the learning process can improve the representation learning of the base model. This improved representation better protect against forgetting of the base tasks when adapting to novel tasks.

Weighted Distillation. The result of the first stage classifier using a cross-entropy loss, as defined in the section before, is $p = softmax(z) \in R^{K_1}$, where z is the set of logits. The classification loss in the first training stage is defined as:

$$L_C(y,p) = -\sum_{i=1}^{K_1} y_i \cdot \log(p_i) \tag{1}$$

Where p_i is the predicted probability scores for each of the classes in the new task, y_i the corresponding ground truth in a one-hot encoding form. In subsequent sessions, a knowledge distillation term is used in the objective, to allow inclusion of past knowledge in the optimization process ($\hat{y}$ are the final layer class probability scores for the new task classes prior to the softmax operation):

$$L_D(z_{old}, \hat{y}) = -\sum_{i=1}^{N} softmax(\frac{z_{old}}{T}) \cdot \log(softmax(\frac{\hat{y}_i}{T})) \tag{2}$$

The logits and predictions are softened in a distillation setting with a temperature term *T*. Softening helps create a smoother transition between the probability scores in the logits set as after a cross-entropy based optimization. This is addressed by spreading out the probability distribution scores with the temperature term. Here, z_{old} represents the logits from the past task, with class-specific average logits computed to obtain a sum of class-weighted logits as:

$$z_{old} = \sum_{i=1}^{K_1} w_i z_i \tag{3}$$

The logits of individual classes z_i, $i \in [|1, K_1|]$ are obtained by averaging pre-softmax scores (with sigmoid activation) for exemplars from the K_1 classes. The summation weights $(w_1, w_2, \dots, w_{k_1})$ are calculated as inverse of class-specific AUC on validation data of Stage 1 classes. This boosts the importance of logits from a more difficult or underrepresented class (lower the class AUC, higher the class weight) in its contribution to the overall representation to be retained for Stage 1 learning. An initial imbalance of classes is propagated upon a distillation based regularization for old classes, exacerbating the overall imbalance when not retaining Stage 1 exemplars for mixing with Stage 2 data. Current methods to mitigate the influence of imbalanced classes like augmentation, weighted cross-entropy etc. are designed for cases when all classes to be learnt have data available unlike in distillation when samples from initial classes are not retained for incremental training. Then the overall objective is (λ set at 0.5 by grid search):

$$L = \lambda L_C + (1 - \lambda)L_D \tag{4}$$

2.1 Training

The Stage 1 task uses data from [28]. Fetal ultrasound videos were acquired with a simultaneous recording of sonographer eye-tracking data. Recording and storage was in compliance with local data governance policies. This data comprised 23016 abdomen, 24508 head, 12839 femur frames. A Tobii Eye Tracker 4C (Tobii, Sweden) recorded point-of-gaze as relative (x,y) coordinates with timestamps 90 Hz capturing 3 gaze points per frame. Gaze points that were less than $0.5°$ apart were combined to a single fixation point. A sonographer visual attention map was generated for each frame using a truncated Gaussian with width equivalent to visual angle of $0.5°$ around the fixation point. In the incremental session, the data comprised fetal cardiac viewing planes with 9386 frames from the 4CH, 6780 frames of LVOT and 6210 of 3VV views. So we use two datasets *D1* and *D2* in initial (Stage 1) and incremental session (Stage 2) respectively.

Model. We design the twin input base model to have parallel convolutional processing strands through the course of residual blocks, to allow for independent convolutional operations on the image and the associated gaze map. The strands are derived from a ResNet-50 architecture (ablations with other backbones in Fig. 2) and configured to accept grayscale inputs. After the final residual blocks, flattened feature maps of both the strands are concatenated. The inclusion of gaze maps follows the protocol in [23]. The fused layer feeds into a fully-connected layer of 512 units, followed by a pre-softmax layer with a sigmoid activation to obtain classwise probabilities. It is a departure from standard ResNet models where an average pooling layer succeeds the final residual blocks and feeds to a dense layer of dimensions equal to the number of classes. This allows to aggregate features and consider the impact of gaze incorporation on performance.

Training. For both stages and their datasets, data augmentation was performed with a 20° rotational augmentation and horizontal flips for the image and gaze map frames [17]. Ultrasound frames and associated gaze maps were resized to 224 × 224. Models were trained with a subject-wise 80:20 split (71 subjects for train and 18 for test) of the dataset. Stage 1 models were trained for 200 epochs with learning rate of 0.001 and adaptive moment estimation (Adam) [12]. Stage 2 models were trained for 200 epochs over the *(N, label, logit)* set for all N frames passed to the trained model. The softening temperature was set at 3.0 after a grid search in $T \in [|1,5|]$. The study is labeled as **Gaze Dist (wt)** when weighted distillation is used. If distillation is used without weights determined by initial task results, it is **Gaze Dist** else it is **No Gaze Dist**. Here, incremental task still uses with weighted distillation. For transfer learning benchmarks, the case with gaze maps available is labeled **FT (gaze)**, and **FT (no gaze)** otherwise. The incremental stage is designed to be able to accept both the image-level inputs as frames alone to make it resilient to cases of unavailable or corrupt gaze data. Our models use 48.7 million parameters, with average training time per epoch of 149 s on a cluster of 2 24 GB Nvidia K80 GPUs. Gaze processing was done on a desktop computer with a 3.1 GHz Intel Core i7 8th generation processor with 512 MB RAM. Models were implemented in Tensorflow 2.0 with eager execution.

3   Results and Discussion

Metrics. We report classwise and average accuracies for Stage 1 and 2 in Table 1, and AUCs in Table 2 for initial classes. AUCs are reported per class as 1-vs-all values and averaged for each stage. To track forgetting, differences in average 1-vs-all AUC values over incremental stages are reported for old classes as *AUCD-iff*, along with fall in accuracy (*AccDiff*). Adaptation to new tasks is directly noted by accuracy and AUC values on Stage 2 classes (Table 3), as this stage builds on Stage 1 learning, with both forgetting and transfer effects. Forward transfer effects are implicit in the new task accuracies and AUC metrics. The relative decline in accuracy and AUC metrics for the initial task, across Stages 1 and 2, encodes combined forgetting and backward transfer effects.

Table 1. Stage 1 and Stage 2 class-specific accuracies for initial task classes, and averages, *AccDiff* quantifies the average drop in accuracy for the old classes

	Stage 1				Stage 2				*AccDiff*
	AC	HC	FL	Avg	AC	HC	FL	Avg	Δ Avg
Gaze Dist (wt)	0.91	0.88	0.87	0.89	0.86	0.85	0.83	0.85	**0.04**
No Gaze Dist	0.73	0.76	0.74	0.74	0.65	0.63	0.62	0.63	0.11
Gaze Dist	0.91	0.88	0.87	0.89	0.83	0.80	0.79	0.81	0.08
FT (gaze)	0.91	0.88	0.87	0.89	0.63	0.54	0.56	0.58	0.31
FT (no gaze)	0.73	0.76	0.74	0.74	0.55	0.49	0.47	0.50	0.24
Lwf/ewc	0.74	0.76	0.74	0.75	0.60	0.61	0.59	0.60	0.15
LwM	0.73	0.76	0.75	0.75	0.64	0.62	0.61	0.62	0.12
PDR	0.73	0.76	0.75	0.75	0.61	0.58	0.57	0.59	0.16
DDE	0.73	0.76	0.75	0.75	0.66	0.63	0.65	0.65	0.10

Results. Classwise performance for the old task is reported for the initial and the incremental sessions in terms of accuracy and 1-vs-all AUC. The change in performance is the difference in these values across sessions. Overall accuracies and averaged AUC are reported for all classes seen until a given session to capture overall model performance (Table 1). Effects of gaze in reducing forgetting are evident as the difference in class-specific AUC is reduced compared to cases not using gaze maps as additional input. We do not retain past exemplars in memory for continual learning and incremental regularization is solely by saved logits from Stage 1 when optimizing for Stage 2 classes. Unlike existing approaches selectively retaining past data, we prioritize reduction in memory footprints while attaining superior continual learning performance using expert insights.

Discussion. Inclusion of gaze implies additional parameters and an incremental computational budget compared to off-the-shelf baselines. For fair comparison, we modified baseline to keep the number of parameters in the same order of magnitude as our proposed multimodal pipeline: 1) For baselines that do not use gaze maps, a parallel set of convolutional layers accepts the image as a redundant input, so the computational budget is comparable to a gaze based approach; 2) For external baselines from literature, we modified baseline representation learning stages to have parallel strands of convolutional layers as before, and enabled them to accept paired inputs, extending for comparison to cases of paired images and gaze maps used as inputs. We choose baselines suitable for contextualizing both weighted distillation and human knowledge inclusion. Comparisons are performed with adapted versions of these methods– LwF.EWC [11] proposed for X-ray incremental learning, distillation and retrospection (PDR) [9], dual distillation and ensembling (DDE) [15] and Learning without Memorizing (LwM) [4] (LwM does not retain exemplars). Methods using distillation outperform transfer learning baselines in terms of knowledge retention evident from higher Stage 2 accuracy and 1-vs-all AUC values. Gaze-based models perform better on Stage 2 metrics as well. Some methods like LwF.MC [14] and iCaRL [30] are equivalent to studies without gaze maps ('**No Gaze Dist**' in Tables) and are not separately benchmarked. Superior results for gaze-driven methods show that additional modalities enable deep networks to learn better input representations. Softening partly smoothens incorrect labels in input spaces for old tasks, reducing forward propagation of inaccuracies. Gains for complex classes such as fetal head frames are notable when using gaze. Unweighted distillation with gaze scales poorly compared to weighted distillation, underlining the role of more complex classes as forgetting is more prominent for a class with intraclass variations or difficult examples due to artefacts like shadows, speckle etc. [18,27]. An weighting strategy informed by initial performance metrics is seen to reduce forgetting here. Specific to accuracies and AUCs for finetuning baselines, gaze inclusion seems to cause slightly higher forgetting than otherwise. This is potentially due to finetuning being carried out across the parameter space and multimodal data causing stronger representation learning in old and new tasks, leading to greater shifts in magnitudes of parameters if no efforts are made to reduce forgetting. Ablations with different CNN backbones (Fig. 2) show that ResNet-50 based models outperform other backbones.

Table 2. Stage 1 and Stage 2 class-specific 1-vs-all AUC, and averages, *AUCdiff* quantifies the average drop in 1-vs-all AUC for the old classes.

	Stage 1				Stage 2				*AUCdiff*
	AC	HC	FL	Avg	AC	HC	FL	Avg	ΔAvg
Gaze Dist (wt)	0.96	0.93	0.89	0.93	0.91	0.85	0.83	0.86	**0.06**
No Gaze Dist	0.75	0.77	0.72	0.75	0.63	0.62	0.61	0.62	0.13
Gaze Dist	0.9	0.89	0.86	0.88	0.81	0.78	0.79	0.79	0.09
FT (gaze)	0.9	0.89	0.86	0.88	0.65	0.6	0.58	0.61	0.27
FT (no gaze)	0.75	0.77	0.72	0.75	0.57	0.52	0.51	0.53	0.21
Lwf/ewc	0.75	0.79	0.73	0.76	0.61	0.63	0.60	0.61	0.15
LwM	0.75	0.79	0.73	0.76	0.64	0.61	0.61	0.62	0.14
PDR	0.75	0.79	0.73	0.76	0.63	0.65	0.62	0.63	0.13
DDE	0.75	0.79	0.73	0.76	0.68	0.67	0.64	0.66	0.10

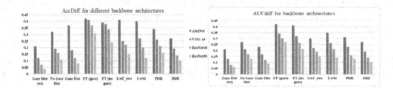

Fig. 2. The variation of *AccDiff* and *AUCdiff* metrics when different backbones are used for learning the fused representations. ResNet50 based pipelines show the least *AccDiff* and *AUCdiff* consistently across studied approaches and adapted baselines.

Table 3. Stage 2 class-specific accuracies and 1-vs-all AUC for new task classes (the avg accuracy and AUC are a proxy for adaptation to the new task)

	Accuracy				1-vs-all AUC			
	4C	3V	LVOT	Avg	4C	3V	LVOT	Avg
Gaze Dist (wt)	0.89	0.8	0.82	0.84	0.92	0.81	0.76	0.83
No Gaze Dist	0.85	0.77	0.74	0.79	0.87	0.76	0.73	0.79
Gaze Dist	0.87	0.78	0.8	0.82	0.85	0.77	0.82	0.81
FT (gaze)	0.86	0.75	0.78	0.80	0.87	0.73	0.79	0.80
FT (no gaze)	0.83	0.72	0.73	0.76	0.82	0.75	0.71	0.76
Lwf/ewc	0.80	0.67	0.66	0.71	0.80	0.69	0.65	0.71
LwM	0.81	0.70	0.68	0.73	0.80	0.73	0.71	0.75
PDR	0.80	0.69	0.67	0.72	0.81	0.72	0.69	0.74
DDE	0.82	0.71	0.67	0.73	0.80	0.70	0.73	0.74

4   Conclusion

We proposed a multimodal pipeline for incremental learning in ultrasound imaging using sonographer eye-tracking data. The inclusion of gaze priors reduced forgetting and enabled performance gains over state-of-the-art methods without requiring retention of past tasks' data. Further, we developed an weighted logits approach for regularization of future task learning, and conceptualized new metrics to assess forgetting and new task adaptation.

References

1. Cai, Y., Sharma, H., Chatelain, P., Noble, J.A.: Sonoeyenet: Standardized fetal ultrasound plane detection informed by eye tracking. In: 2018 IEEE 15th International Symposium on Biomedical Imaging (ISBI 2018), pp. 1475–1478 (2018)
2. Chen, H., Ni, D., Yang, X., Li, S., Heng, P.A.: Fetal abdominal standard plane localization through representation learning with knowledge transfer. In: Wu, G., Zhang, D., Zhou, L. (eds.) MLMI 2014. LNCS, vol. 8679, pp. 125–132. Springer, Cham (2014). https://doi.org/10.1007/978-3-319-10581-9_16
3. Davis, J., Goadrich, M.: The relationship between precision-recall and roc curves. In: Proceedings of the 23rd International Conference on Machine Learning. p. 233–240. ICML 2006, Association for Computing Machinery, New York, NY, USA (2006). https://doi.org/10.1145/1143844.1143874
4. Dhar, P., Singh, R.V., Peng, K.C., Wu, Z., Chellappa, R.: Learning without memorizing. In: Proceedings of the IEEE Conference on Computer Vision and Pattern Recognition, pp. 5138–5146 (2019)
5. Díaz-Rodríguez, N., Lomonaco, V., Filliat, D., Maltoni, D.: Don't forget, there is more than forgetting: new metrics for continual learning. arXiv:1810.13166 (2018)
6. Droste, R., et al.: Ultrasound image representation learning by modeling sonographer visual attention. In: Chung, A.C.S., Gee, J.C., Yushkevich, P.A., Bao, S. (eds.) IPMI 2019. LNCS, vol. 11492, pp. 592–604. Springer, Cham (2019). https://doi.org/10.1007/978-3-030-20351-1_46
7. Goodfellow, I.J., Mirza, M., Xiao, D., Courville, A., Bengio, Y.: An empirical investigation of catastrophic forgetting in gradient-based neural networks. arXiv:1312.6211 (2013)
8. Hinton, G., Vinyals, O., Dean, J.: Distilling the knowledge in a neural network. In: NIPS 2014 Deep Learning Workshop (2014)
9. Hou, S., Pan, X., Change Loy, C., Wang, Z., Lin, D.: Lifelong learning via progressive distillation and retrospection. In: ECCV (2018)
10. Kemker, R., Kanan, C.: Fearnet: Brain-inspired model for incremental learning. arXiv:1711.10563 (2017)
11. Kim, H.-E., Kim, S., Lee, J.: Keep and learn: continual learning by constraining the latent space for knowledge preservation in neural networks. In: Frangi, A.F., Schnabel, J.A., Davatzikos, C., Alberola-López, C., Fichtinger, G. (eds.) MICCAI 2018. LNCS, vol. 11070, pp. 520–528. Springer, Cham (2018). https://doi.org/10.1007/978-3-030-00928-1_59
12. Kingma, D.P., Adam, J.B.: A method for stochastic optimization. arXiv:1412.6980
13. Kirkpatrick, J., et al.: Overcoming catastrophic forgetting in neural networks. In: Proceedings of the National Academy of Sciences, vol. 114, no. 13, pp. 3521–3526 (2017)

14. Li, Z., Hoiem, D.: Learning without forgetting. IEEE Trans. Pattern Anal. Mach. Intell. **40**(12), 2935–2947 (2017)
15. Li, Z., Zhong, C., Wang, R., Zheng, W.-S.: Continual learning of new diseases with dual distillation and ensemble strategy. In: Martel, A.L., et al. (eds.) MICCAI 2020. LNCS, vol. 12261, pp. 169–178. Springer, Cham (2020). https://doi.org/10.1007/978-3-030-59710-8_17
16. Omar, H., Patra, A., Domingos, J., Upton, R., Leeson, P., Noble, J.: Myocardial wall motion assessment in stress echocardiography by quantification of principal strain bulls eye maps: P299. Eur. Heart J. Cardiovascular Imaging **18** (2017)
17. Omar, H.A., Domingos, J.S., Patra, A., Leeson, P., Noble, J.A.: Improving visual detection of wall motion abnormality with echocardiographic image enhancing methods. In: 2018 40th Annual International Conference of the IEEE Engineering in Medicine and Biology Society (EMBC), pp. 1128–1131. IEEE (2018)
18. Omar, H.A., Domingos, J.S., Patra, A., Upton, R., Leeson, P., Noble, J.A.: Quantification of cardiac bull's-eye map based on principal strain analysis for myocardial wall motion assessment in stress echocardiography. In: 2018 IEEE 15th International Symposium on Biomedical Imaging (ISBI 2018), pp. 1195–1198. IEEE (2018)
19. Omar, H.A., Patra, A., Domingos, J.S., Leeson, P., Noblel, A.J.: Automated myocardial wall motion classification using handcrafted features vs a deep CNN-based mapping. In: 2018 40th Annual International Conference of the IEEE Engineering in Medicine and Biology Society (EMBC), pp. 3140–3143. IEEE (2018)
20. Ozdemir, F., Fuernstahl, P., Goksel, O.: Learn the new, keep the old: extending pretrained models with new anatomy and images. In: Frangi, A.F., Schnabel, J.A., Davatzikos, C., Alberola-López, C., Fichtinger, G. (eds.) MICCAI 2018. LNCS, vol. 11073, pp. 361–369. Springer, Cham (2018). https://doi.org/10.1007/978-3-030-00937-3_42
21. Patra, A., Huang, W., Noble, J.A.: Learning spatio-temporal aggregation for fetal heart analysis in ultrasound video. In: Cardoso, M.J., et al. (eds.) DLMIA/ML-CDS-2017. LNCS, vol. 10553, pp. 276–284. Springer, Cham (2017). https://doi.org/10.1007/978-3-319-67558-9_32
22. Patra, A., et al.: Sequential anatomy localization in fetal echocardiography videos. arXiv preprint arXiv:1810.11868 (2018)
23. Patra, A., et al.: Efficient ultrasound image analysis models with sonographer gaze assisted distillation. In: Shen, D., et al. (eds.) MICCAI 2019. LNCS, vol. 11767, pp. 394–402. Springer, Cham (2019). https://doi.org/10.1007/978-3-030-32251-9_43
24. Patra, A., Chakraborti, T.: Learn more, forget less: Cues from human brain. In: Proceedings of the Asian Conference on Computer Vision (ACCV), November 2020
25. Patra, A., Noble, J.A.: Incremental learning of fetal heart anatomies using interpretable saliency maps. In: Zheng, Y., Williams, B.M., Chen, K. (eds.) MIUA 2019. CCIS, vol. 1065, pp. 129–141. Springer, Cham (2020). https://doi.org/10.1007/978-3-030-39343-4_11
26. Patra, A., Noble, J.A.: Multi-anatomy localization in fetal echocardiography videos. In: 2019 IEEE 16th International Symposium on Biomedical Imaging (ISBI 2019), pp. 1761–1764. IEEE (2019)
27. Patra, A., Noble, J.A.: Hierarchical class incremental learning of anatomical structures in fetal echocardiography videos. IEEE J. Biomed. Health Informatics (2020)
28. PULSE: Perception ultrasound by learning sonographic experience (2018). www.eng.ox.ac.uk/pulse
29. Ravishankar, H., et al.: Understanding the mechanisms of deep transfer learning for medical images. In: Deep Learning and Data Labeling for Medical Applications, pp. 188–196. Springer, Cham (2016). https://doi.org/10.1007/978-3-319-46976-8_20

30. Rebuffi, S.A., Kolesnikov, A., Sperl, G., Lampert, C.H.: ICARL: Incremental classifier and representation learning. In: Proceedings of the IEEE CVPR, pp. 2001–2010 (2017)
31. Rusu, A.A., et al.: Progressive neural networks. arXiv:1606.04671 (2016)
32. Zhang, J., Wang, Y.: Continually modeling Alzheimer's disease progression via deep multi-order preserving weight consolidation. In: Shen, D., et al. (eds.) MICCAI 2019. LNCS, vol. 11765, pp. 850–859. Springer, Cham (2019). https://doi.org/10.1007/978-3-030-32245-8_94

Development and Evaluation of Intraoperative Ultrasound Segmentation with Negative Image Frames and Multiple Observer Labels

Liam F. Chalcroft[1,2]([✉]), Jiongqi Qu[1], Sophie A. Martin[1,3,4],
Iani JMB Gayo[1,3,5], Giulio V. Minore[6], Imraj RD Singh[7],
Shaheer U. Saeed[1,3,5], Qianye Yang[1,3,5], Zachary M. C. Baum[1,3,5],
Andre Altmann[1,3], and Yipeng Hu[1,3,5]

[1] Department of Medical Physics and Biomedical Engineering,
University College London, London, UK
liam.chalcroft.20@ucl.ac.uk
[2] Wellcome Center for Human Neuroimaging, University College London,
London, UK
[3] Centre for Medical Image Computing, University College London, London, UK
[4] Dementia Research Centre, UCL Institute of Neurology,
University College London, London, UK
[5] Wellcome/EPSRC Centre for Interventional and Surgical Sciences,
University College London, London, UK
[6] Department of Physics and Astronomy, University College London, London, UK
[7] Department of Computer Science, University College London, London, UK

Abstract. When developing deep neural networks for segmenting intra-operative ultrasound images, several practical issues are encountered frequently, such as the presence of ultrasound frames that do not contain regions of interest and the high variance in ground-truth labels. In this study, we evaluate the utility of a pre-screening classification network prior to the segmentation network. Experimental results demonstrate that such a classifier, minimising frame classification errors, was able to directly impact the number of false positive and false negative frames. Importantly, the segmentation accuracy on the classifier-selected frames, that would be segmented, remains comparable to or better than those from standalone segmentation networks. Interestingly, the efficacy of the pre-screening classifier was affected by the sampling methods for training labels from multiple observers, a seemingly independent problem. We show experimentally that a previously proposed approach, combining random sampling and consensus labels, may need to be adapted to perform well in our application. Furthermore, this work aims to share practical experience in developing a machine learning application that assists highly variable interventional imaging for prostate cancer patients, to present robust and reproducible open-source implementations, and to

L. F. Chalcroft, J. Qu, S. A. Martin, I. J. M. B. Gayo, G. V. Minore, I. R. D. Singh—
Contributed equally.

J. A. Noble et al. (Eds.): ASMUS 2021, LNCS 12967, pp. 25–34, 2021.
https://doi.org/10.1007/978-3-030-87583-1_3

report a set of comprehensive results and analysis comparing these practical, yet important, options in a real-world clinical application.

1   Introduction

Many urological procedures for prostate cancer patients, such as ablation therapy and needle biopsies, are guided by B-mode transrectal ultrasound images (TRUS) to identify and then monitor the shape and location of prostate glands [2,11]. This application is useful for a number of interventional tasks, such as estimating the gland size, regions of pathological interest and surrounding healthy, but vulnerable, tissues. However, due to variable acoustic coupling, inhomogeneous intensity distribution, and the necessity of real-time monitoring, delineating the boundaries of prostate glands is a challenging task, even for experienced urologists. Deep neural networks have been proposed to automate this process [1,5,7,8,14].

The performance of these networks relies on well-defined ground truth labels. To date, there is no gold standard approach in many ultrasound imaging applications with high inter-and intra-rater variability in labelling and its use in training. Existing approaches deal with multiple labels by using a pixel-level voting strategy or random sampling, both estimating the expected labels. In [12], Sudre et al. observed that combining random and voting strategies during training improves stability and performance in the context of brain lesion detection. In this paper, we consider labels from multiple independent raters and investigate the effect of different sampling strategies during segmentation, and test these proposed sampling methods in the context of interventional ultrasound imaging for prostate cancer patients.

In addition to the label variability, ultrasound data itself is known to be of high variance, due to its user dependency and flexible use protocols. For example, it is common that some frames do not contain the region of interest (ROI), particularly due to the small size of the prostate gland in our application. The presence of negative frames presents a key challenge in segmentation, as wrongfully segmenting a frame that does not contain the ROI could potentially lead to misdiagnosis or damage to healthy tissues. Using a widely-used segmentation accuracy metric based on overlap, such as Dice, to quantify this error can be problematic. The naive implementation of Dice is independent of the number of false positive pixels and the cost of negative frames may not be easily quantified with respect to the cost of negative pixels when designing a new loss function. For example, in the case of a handheld setting, relative positions and distances in the out-of-plane direction between ultrasound frames are in general variable and unknown, which may lead to an unspecified misjudgement of where the ROI boundary is, given a false positive frame. A separate frame classification may provide more intuitive user guidance when using the segmentation algorithms.

Limited work has been proposed to address the problem of negative frames within medical image segmentation. In [3], false positives in a video object segmentation task were reduced through the introduction of a post-processing classifier. In [10], meta-classification was used to detect false positive samples in semantic segmentation. This has motivated a screening strategy in this work that can detect negative frames before they are incorrectly segmented by the segmentation network. Such a separately trained classification network can also provide flexible control at test-time between false positive and false negative rates on a frame-level, which is arguably more difficult to achieve by altering threshold on pixel-level class probability in a segmentation network. Alternative approaches and different loss functions to address this issue are also discussed or compared in this paper.

2   Methods

2.1   Segmentation Network

U-Net [9], a fully convolutional neural network, is adapted from a well-established reference implementation. Our network consists of 5 layers that starting with initial 16 channels, with residual network blocks replacing the original individual convolutional layers to encourage fast convergence [4]. Images were normalised to zero-mean and unit-variance. All the segmentation networks were trained with a mini-batch size of 32, using the Adam optimiser [6] with an exponential learning rate scheduler that minimises a soft Dice loss function: $L_{SoftDice} = \frac{2\Sigma y_{pred} \cdot y_{true}}{\Sigma y_{pred} + \Sigma y_{true}}$, where Σ is the pixel-wise sum, y_{pred} is the predicted class probabilities and y_{true} is the ground truth mask. The Dice value was also used to monitor validation set performance. Random data augmentations are applied during training with probability $p = 0.3$, including random affine deformations (rotation $|\theta_r| \leq 2.5 \deg$, maximum translation 0.05, scaling in range $0.95-1$), and random flipping along the vertical axis. These augmentations were empirically found robust for the TRUS data in this application.

2.2   Frame Classification Network

A reference-quality ResNeXt [15] classifier pre-trained on ImageNet was adapted to predict whether a prostate is present based on the frame-level consensus. The network was modified to accept single channel and resized to 224×224. The weights are normalised with mean and standard deviation (0.449, 0.226), representing the average of the three original RGB channels. This model was trained with an initial learning rate of 0.0001, using the Adam optimiser and a binary cross-entropy loss function.

2.3   Label Sampling

Six different label sampling methods were investigated and evaluation results on the hold-out test data are reported. The methods are summarised in Table 1. The combination label strategy randomly selects a certain percentage of the data to perform the vote sampling method and applies the random sampling method to the remainder data.

Table 1. Summary of label sampling methods. Soft mean refers to the non-rounded mean of labels, treated as a continuous probability map.

Label strategy	Description
Vote	Pixel-level majority voting from the 3 labels
Random	Single label selected at random
Mean	Soft mean of the 3 labels
Combination (25%)	Combination of 25% vote and 75% random labels
Combination (50%)	Combination of 50% vote and 50% random labels
Combination (75%)	Combination of 75% vote and 25% random labels

2.4   Pre-screening Strategy

The classifier can be combined with the segmentation network to facilitate a pre-screening strategy illustrated in Fig. 1d. The frame will pass to the segmentation network only if the classifier-predicted probability is greater than a set threshold in logits, whose values from 0 to 5 are tested based on observations of resulting classification accuracy range on the validation set. Different ways of combining the classification and segmentation networks is also possible and remains interesting for future investigation.

2.5   Loss Functions for Segmentation

For a given label sampling method, we test different segmentation loss functions. This allows us to ascertain whether the frame-level classification can also be handled by the segmentation directly, as opposed to the above-described pre-screening. Two alternatives are considered in addition to the Dice loss function, a combo loss with an equal weighting between dice loss and binary cross entropy loss (BCE), and a weighted binary cross entropy loss based on [13] (W-BCE). The equation for the Dice-BCE loss is given by:

$$Dice\text{-}BCE = 0.5 \times (1 - Dice) + 0.5 \times BCE \tag{1}$$

where the binary cross entropy loss is defined by:

$$BCE = -\sum_{n}^{N} x_n \log p_n + (1 - x_n) \log (1 - p_n) \tag{2}$$

where N is the number of pixels, x_n is target class per pixel and p_n is the predicted probability from the network. The BCE loss can be modified to assign weights, w_c to each class (c = 0, 1) such that:

$$W\text{-}BCE = -\sum_{n}^{N} w_0 x_n \log p_n + w_1 (1 - x_n) \log (1 - p_n) \tag{3}$$

where in our case, $w_1 = \frac{1}{\sum^{N} x_n = 1}$ and $w_0 = 1 - w_1$.

2.6   Evaluation Experiments

The Dice coefficient is computed on positive frames excluding those that are predicted to be negative by the segmentation network or by, when in use, the pre-screening classifier, to ensure that we do not penalise the network for correctly identifying negative frames (a 0 Dice coefficient). In addition, we report frame-level classification performance for both frame classifier and segmentation network, when the latter is used without pre-screening. In this case, rates of false positive frames and their false positive area are computed. All results are reported on the independent hold-out test set. *p-values* from t-tests at significance level of 0.05 are also reported when comparison is made.

The dataset used in this study contains 2D B-mode transrectal ultrasound frames from 250 patients. For each subject, a range of 50–120 frames were acquired at the start of the procedure, with a bi-plane transperineal ultrasound probe (C41L47RP, HI-VISION Preirus, Hitachi Medical Systems Europe) and a digital transperineal stepper (D&K Technologies GmbH, Barum, Germany) to view and scan entire gland. For labelling, 6644 ultrasound images were sampled with size 403×361 and were manually annotated by three independent raters. A set of example frames are shown in Fig. 1a–1c with varying label agreement.

At the patient level, 5224 and 1346 frames were sampled for training/validation and hold-out test, an 80:20 split. The networks were trained using a 3-fold cross-validation ensemble strategy, with 3484 and 1740 samples for training and validation in each fold, respectively. Predictions from each of the networks were averaged at test-time to generate a single probability map that is converted into a mask during inference on the hold-out set. The code is made publicly available at https://github.com/sophmrtn/RectAngle.

3   Results and Discussion

Label Sampling. The performance of the segmentation network for each sampling method is shown via box plots in Fig. 2a. The mean label sampling strategy was statistically different (all $p-values < 0.05$) from all other methods. All other sampling methods obtained similar performance.

The pre-screening classifier achieved an accuracy of 97.1% on the validation dataset during training. Table 2 summarises the Dice values with and without the pre-screening for the six label sampling methods. The classifier is shown to improve performance significantly for the mean label strategy ($p = 0.001$).

Classification Threshold. The threshold used by the classifier plays a role in controlling the false positive frame rate seen by the segmentation network and can therefore be tuned as a variable at test time. We therefore tested a range of thresholds from 0 to 5 corresponding to probabilities of 0.5 to 1 and observe the effect on the mean Dice for each label sampling method. This is shown in Fig. 2b. From this plot the combination of consensus and random labels with a ratio of 25% and 75% respectively leads to the highest Dice score and this increases with threshold in general for all label sampling methods.

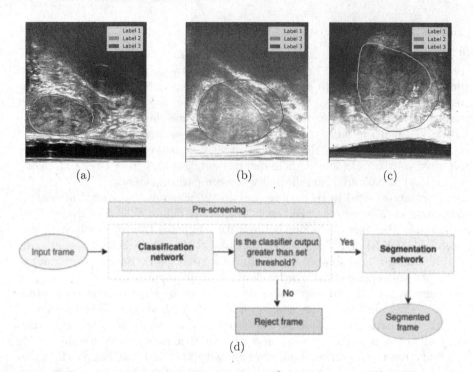

Fig. 1. a–c) Example frames are shown with manual labels from three observers in green, red and blue respectively. a) All labels are in close agreement. b) Two labellers agree however one annotation is significantly larger. c) Only two labellers identify the prostate presence, but with slightly different locations. d) Flowchart to describe the pre-screening strategy. (Color figure online)

Table 2. The Dice coefficient values on the hold-out test data (mean ± std. dev.) with and without pre-screening. The median values are reported for inspecting skewness. Statistically significant improvement ($p < 0.05$) are in bold.

Sampling method	Mean dice			Median dice	
	w	w/o	p-val	w	w/o
Vote	0.866 ± 0.180	0.856 ± 0.197	0.220	0.927	0.926
Random	0.867 ± 0.184	0.857 ± 0.200	0.223	0.926	0.925
Mean	**0.861 ± 0.184**	**0.831 ± 0.236**	**0.001**	0.920	0.917
Combine (25%)	0.866 ± 0.182	0.857 ± 0.198	0.273	0.926	0.925
Combine (50%)	0.867 ± 0.180	0.859 ± 0.197	0.328	0.927	0.927
Combine (75%)	0.870 ± 0.174	0.861 ± 0.190	0.253	0.926	0.925

Pre-screening Classifier. The pre-screened segmentation model can be used to examine the effect on the number of false positives/negatives on both frame and pixel levels. We also use the modified loss functions to compare the

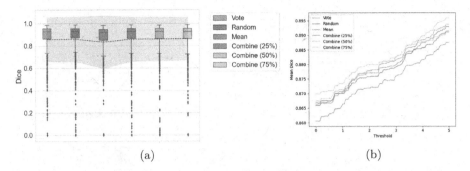

Fig. 2. a) Dice coefficients for positive predictions on hold-out set of 1346 frames. Dashed line shows mean Dice values for each strategy, with shading indicating the standard deviation from the mean. b) The mean Dice score for positive frames is reported for a range of classification thresholds for each label sampling method during segmentation. The standard deviation is omitted in the figure for readability, where for the combination strategy (25%) we obtain a standard deviation of 0.15 at a threshold of 5.

performance of the segmentation network alone with a loss chosen to tackle both tasks simultaneously. The FPR and FNR is computed for the different labelling strategies in each case as shown in Fig. 3a. From these results we observe a slight decrease in the number of false positive frames as the threshold increases. The most noticeable effect of threshold is on the FNR for which a larger threshold leads to a greater number of false negative frames. On the other hand, the loss function is shown to be effective to some extent at addressing the frame-level classification task. The losses seemed to lead to a lower false negatives than false positive frames, although altering the weight to control the two type of frame-level errors does not seem to be straightforward. This is consistent with what can be observed from the areas of the false segmentations.

Further inspecting the labels from three observers overlaid with those frames, on which the segmentation and classifier networks disagreed for the frame classification, as shown in the examples in Fig. 3b, c. Interestingly, relatively large disagreement between observers can also be found on those network-disagreed images. This may suggest a correlation between the label sampling methods and the frame classifying strategy. This is also supported by the results in Table 2, where, for example, highest median Dice values may come from different label sampling methods, between models with and without the pre-screening strategy.

This paper reports experiment results with and without an independently-trained pre-screening classifier. Future work may investigate a classifier trained simultaneously with the segmentation network, such that the segmentation network could be optimised on representative frames that need to be segmented.

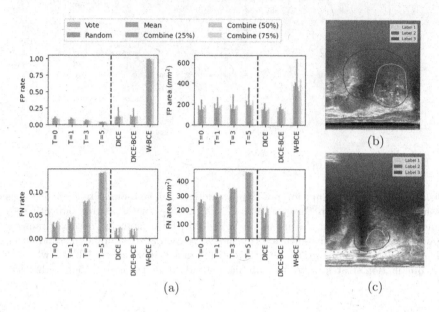

Fig. 3. a) False positive (FP) and false negative (FN) rates (frame-level) and areas (pixel-level) are computed for each label sampling method using different screening thresholds. We also show the rates achieved using different loss functions; Dice, Dice-BCE and W-BCE, using the segmentation-only approach (Dotted black line used to separate these cases, with classifier used in results left of line and segmentation-only results to the right) (b–c) Example frames with manual labels, where the classifier and segmentation network disagreed. b) Classifier predicted the presence of prostate, but segmented mask is empty. c) Classifier predicted an empty frame, but the prostate was segmented. In both cases, only two labellers were in agreement, but not over the size and position of the prostate.

4   Conclusion

In this study, we investigated different strategies for handling multiple labels for intraoperative prostate gland segmentation on TRUS images. We demonstrate that disagreements between labellers affect the performance of a U-Net segmentation network due to the difficulty when defining a ground truth. Whilst there were no significant differences between the label sampling methods themselves using the Dice loss, by introducing a pre-screening strategy with a separate classifier, we show an improved segmentation accuracy by removing false positive frames. This was observed for the mean label strategy ($p = 0.001 < 0.05$) between the mean Dice with, and without, pre-screening. Our results also agree in general with existing findings that using a combination of random and consensus labels (25%, 75% respectively) during training leads to better, and more stable performance with a mean Dice of 0.87 ± 0.17. Alternatively, the segmentation network can be trained using loss functions that aim to address the frame-level classification task in parallel with optimising the Dice score. For these models, we find a better ability to handle false negative frames than using a pre-screening classifier. However, the classifier still provides better flexibility to control the frame-level accuracy during test-time. This work illustrates the potential benefit of pre-screening prior to classification during real-time ultrasound-guided procedures where the reduction of a specific error type may be more desirable.

Acknowledgement. This work is supported by the EPSRC-funded UCL Centre for Doctoral Training in Intelligent, Integrated Imaging in Healthcare (i4health) (EP/S021930/1) and the Department of Health's NIHR-funded Biomedical Research Centre at University College London Hospitals. Z.M.C. Baum is supported by the Natural Sciences and Engineering Research Council of Canada Postgraduate Scholarships-Doctoral Program, and the University College London Overseas and Graduate Research Scholarships. This work is also supported by the Wellcome/EPSRC Centre for Interventional and Surgical Sciences (203145Z/16/Z).

References

1. Anas, E.M.A., Mousavi, P., Abolmaesumi, P.: A deep learning approach for real time prostate segmentation in freehand ultrasound guided biopsy. Med. Image Anal. **48**, 107–116 (2018). https://doi.org/10.1016/j.media.2018.05.010
2. Ghose, S., Oliver, A., et al.: A survey of prostate segmentation methodologies in ultrasound, magnetic resonance and computed tomography images. Comput. Meth. Programs Biomed. **108**(1), 262–287 (2012). https://doi.org/10.1016/j.cmpb.2012.04.006
3. Giordano, D., Kavasidis, I., et al.: Rejecting false positives in video object segmentation. In: Azzopardi, G., Petkov, N. (eds.) Computer Analysis of Images and Patterns, pp. 100–112. Springer International Publishing, Cham (2015). https://doi.org/10.1007/978-3-319-23192-1_9
4. He, K., Zhang, X., et al.: Deep residual learning for image recognition (2015)
5. Hossain, M.S., Paplinski, A.P., Betts, J.M.: Prostate segmentation from ultrasound images using residual fully convolutional network (2019)
6. Kingma, D.P., Ba, J.: Adam: a method for stochastic optimization (2017)
7. Lei, Y., Tian, S., et al.: Ultrasound prostate segmentation based on multidirectional deeply supervised v-net. Med. Phys. **46**(7), 3194–3206 (2019). https://doi.org/10.1002/mp.13577
8. Orlando, N., Gillies, D.J., et al.: Automatic prostate segmentation using deep learning on clinically diverse 3D transrectal ultrasound images. Med. Phys. **47**(6), 2413–2426 (2020). https://doi.org/10.1002/mp.14134
9. Ronneberger, O., Fischer, P., Brox, T.: U-net: convolutional networks for biomedical image segmentation. CoRR abs/1505.04597 (2015). http://arxiv.org/abs/1505.04597
10. Rottmann, M., Maag, K., Chan, R., Hüger, F., Schlicht, P., Gottschalk, H.: Detection of false positive and false negative samples in semantic segmentation (2019)
11. Sarkar, S., Das, S.: A review of imaging methods for prostate cancer detection. Biomed. Eng. Comput. Biol. **7s1**, BECB–S34255 (2016). https://doi.org/10.4137/becb.s34255
12. Sudre, C.H., Anson, B.G., et al.: Let's agree to disagree: learning highly debatable multirater labelling. CoRR abs/1909.01891 (2019). http://arxiv.org/abs/1909.01891

13. Sudre, C.H., Li, W., Vercauteren, T., Ourselin, S., Jorge Cardoso, M.: Generalised dice overlap as a deep learning loss function for highly unbalanced segmentations. In: Cardoso, M., et al. (eds.) Deep Learning in Medical Image Analysis and Multimodal Learning for Clinical Decision Support. DLMIA 2017, ML-CDS 2017. Lecture Notes in Computer Science, pp. 240–248 (2017). https://doi.org/10.1007/978-3-319-67558-9_28

14. Wang, Y., et al.: Deep attentional features for prostate segmentation in ultrasound. In: Frangi, A.F., Schnabel, J.A., Davatzikos, C., Alberola-López, C., Fichtinger, G. (eds.) MICCAI 2018. LNCS, vol. 11073, pp. 523–530. Springer, Cham (2018). https://doi.org/10.1007/978-3-030-00937-3_60

15. Xie, S., Girshick, R., Dollár, P., Tu, Z., He, K.: Aggregated residual transformations for deep neural networks (2017)

Automatic Tomographic Ultrasound Imaging Sequence Extraction of the Anal Sphincter

Helena Williams[1,2,3(✉)], Laura Cattani[1,4], Tom Vercauteren[2], Jan Deprest[1,4], and Jan D'hooge[3]

[1] Department of Development and Regeneration, Cluster Urogenital Surgery, Biomedical Sciences, KU Leuven, Leuven, Belgium
helena.williams@kuleuven.be
[2] School of Biomedical Engineering and Imaging Sciences, King's College London, London, UK
[3] Department of Cardiovascular Sciences, KU Leuven, Leuven, Belgium
[4] Clinical Department of Obstetrics and Gynaecology, UZ Leuven, Leuven, Belgium

Abstract. Transperineal volumetric ultrasound (TPUS) imaging has become routine practice for diagnosing anorectal dysfunction, a life-challenging pelvic floor dysfunction (PFD). To assess the integrity of the whole length of the anal sphincter from three-dimensional (3D) ultrasound (US) data, sonographers first extract a tomographic US imaging (TUI) sequence from the TPUS recording. TUI sequences consist of eight equally spaced and properly oriented two-dimensional (2D) coronal-view slices of the anal sphincter complex. TUI sequences are visually assessed by a sonographer to diagnose anal sphincter injury. Obtaining TUI sequences is performed manually in clinical practice, which is labour-intensive and requires expert knowledge of pelvic floor anatomy. To the best of our knowledge, this work is the first to report an automatic method to aid this medical imaging acquisition task. We propose a novel, convolutional neural network (CNN) approach for the automatic extraction of the TUI sequences from a TPUS. The method utilises a CNN to segment the external anal sphincter (EAS), and the desired TUI sequences are subsequently extracted after several automatic post-processing steps. The proposed method is evaluated on 30 TPUS recordings and compared against manually acquired gold standard TUI sequences. One expert evaluated the quality of the automatically detected TUI sequences in terms of their clinical acceptability for diagnosis. The automatic method performs with an overall clinical acceptability of 90.00%. The method reduces the time required to extract the anal sphincter complex TUI sequence of a TPUS by 52.36 s and may reduce the need for high-level expertise in anorectal dysfunction analysis.

1 Introduction

PFD includes pelvic organ prolapse, urinary incontinence and anorectal dysfunction, including anal incontinence and obstructive defecation. Obstetric anal

© Springer Nature Switzerland AG 2021
J. A. Noble et al. (Eds.): ASMUS 2021, LNCS 12967, pp. 35–44, 2021.
https://doi.org/10.1007/978-3-030-87583-1_4

sphincter injury is the most common finding in women with anal incontinence in reproductive age. Anal sphincter integrity (or injury) can be assessed with exo-anal (TPUS or introital) or with endo-anal US. Endo-anal is more intrusive, and TPUS showed a substantial correlation with exo-anal with high sensitivity for anal sphincter complex evaluation [6]. TPUS has shown to have similar image quality to introital with lower inter-rater variability [2,6,10]. Therefore, TPUS was used in this study, further details can be found in literature [2,3,6].

Within clinical assessment, sonographers use TUI sequences of the anal sphincter complex to visually assess the integrity of the entire anal sphincter [3,7]. TUI sequences consist of eight equally spaced and properly oriented 2D coronal view slices of the anal sphincter complex. Manual extraction of TUI sequences from a TPUS recording is labour intensive and recognised as a highly skilled task, as the sonographer must manually manipulate a TPUS recording to locate prede-termined locations, based on the cranial termination of the EAS and the caudal termination of the internal anal sphincter (IAS) [2,3,6], as shown in Fig. 1. The quality of TUI extraction is heavily dependent on the sonographer's skill, and sig-nificant inter-observer variability may lead to, in extreme cases, misdiagnosis.

Therefore, we aim to automatically extract the TUI sequences from a TPUS recording, to address the limitations above. In this work, the sonographer would only need to acquire a TPUS recording following a standard acquisition, (i.e. the transperineal probe is placed at the opening to the vagina and perpendicularly to the anal canal) [6]. Our solution aims to speed up assessment for skilled sonographers, and potentially allow non-experts to perform these assessments.

We briefly describe our work in the context of related literature that has pro-posed automated image analysis of pelvic floor structures, such as the levator hiatus [1,5,9] and the puborectalis muscle [11]. Automatic assessment of the leva-tor hiatus [1,5] utilised CNNs and active shape models [9], and performed within inter-observer variability. In other work, an automatic clinical solution was pre-sented for the extraction of a plane of interest used in PFD assessment[12]. The

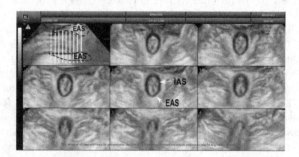

Fig. 1. TUI sequence of a normal anal sphincter. The top left image shows the mid-sagittal plane with the EAS annotated; the eight other images represent coronal slices through the anal canal. The locations of the slices are given by the vertical lines in the midsagittal plane. Slice 1 is the non-dashed vertical line on the left; slice 8 is at the right. The arrows show the location of the EAS and IAS within a coronal view plane.

paper utilised CNN landmark regression, and performed within inter-observer variability, while reducing the time required for assessment by 100 s.

We believe the work presented in this paper is of clinical impact, due to the difficult nature of manipulating TPUS recordings of the anal sphincter, the lack of current automation of TUI extraction, and the expertise required by sonographers. In this paper, we describe to the authors' knowledge the first automatic anal sphincter TUI sequence extraction solution. The proposed solution locates the EAS and extracts eight equidistant 2D images of the anal sphincter in the coronal-view, comparable to manually acquired TUI sequences. This work utilises the advances of CNN segmentation and is evaluated on 30 TPUS recordings. The clinical acceptability and time taken are recorded and compared to an expert sonographer. We believe a fully automatic TUI extraction solution may save clinicians time to allow more focus on patient care and treatment planning.

2   Materials and Methods

During urogynaecological US examination, sonographers aim to evaluate sphincter integrity based on the sonographic appearance of the EAS and IAS. The sonographer acquires a TPUS recording at approximately 60 deg aperture and 70 deg acquisition angle with a 3D convex transducer, when possible during pelvic floor muscle contraction. The TUI sequences are identified in post-processing steps. On the extracted TUI sequences, the sonographer assessed EAS and IAS integrity, and if present measured the degree of tear in the EAS and in the IAS which corresponds to the internationally accepted clinical classification [3]. Before describing the method in detail, we first describe the acquisition protocol.

2.1   Acquisition Protocol

All data was acquired with a Voluson E10 BT16 ultrasound system (GE Healthcare: Zipf, Austria) equipped with a 3D 4–8 MHz convex probe placed transperineally with an average voxel resolution of 0.3 mm by 0.3 mm by 0.3 mm. For testing, a total 30 3D TPUS recordings were acquired. Volumes covering the entire length of the EAS were obtained and post-processed offline on a desktop computer using 4D View Software (GE Healthcare; Austria GmbH & Co, Zipf, Austria) according to the international practice parameter [8].

2.2   The Proposed Pipeline

The proposed method is shown in Fig. 2. Firstly, the EAS was segmented from a TPUS recording, the centre of mass, X_{cm}, was determined and the corresponding mid-sagittal plane extracted. Four parallel planes were extracted and an averaged EAS segmentation was formed. The principal axes of rotation of the averaged segmentation was identified and a rotation matrix was formed. The TPUS was then rotated to ensure the anal sphincter was parallel to the coordinate axes, and eight equidistant slices of the EAS in the coronal view were extracted.

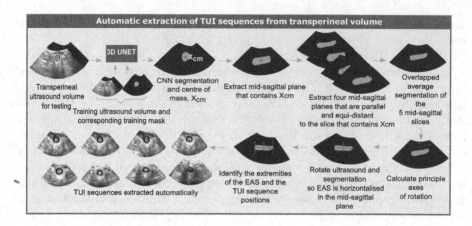

Fig. 2. Proposed pipeline of the automatic TUI extraction algorithm

3D EAS Segmentation. Firstly the EAS was automatically segmented. This was achieved by utilising a CNN, which accepted a TPUS as input and outputted a 3D voxel-wise segmentation of the EAS. The architecture used was 3D U-Net [14], and advanced data augmentation was used including an adaptation of the original mix-up [13], where three images and their labels were linearly combined.

Rotation of the TPUS Recording. During manual acquisition the sonographer may need to rotate the TPUS to horizontalise the anal canal in the mid-sagittal plane. This ensures the axes of rotation of the sphincter lay along the coordinate axes. Here, we describe how the rotation matrix, R, was formed in order to automate this task. Firstly, X_{cm} of the 3D segmentation was identified, and the mid-sagittal plane of the segmentation which contained X_{cm} was extracted. The mid-sagittal plane is given by the x and y directions of the volume data, and is dependent on a standard acquisition protocol used within clinic (i.e. the probe placed at the entrance of the vagina perpendicular to the anal canal). Several equidistant parallel planes to the mid-sagittal plane containing, X_{cm} were extracted and multiplied together to produce an averaged 2D EAS segmentation, based on the common overlap (i.e. common voxel values were equal to 1 and uncommon voxel values were equal to 0). The mid-sagittal planes used contained the coordinate X_{cm}, $X_{cm} \pm 1.5$ mm and $X_{cm} \pm 3$ mm.

Principle component analysis (PCA) was used to identify the eigenvectors, $\vec{v}_{av}$, describing the principle axes of rotation of the averaged 2D EAS segmentation within the mid-sagittal view. PCA was only applied to the mid-sagittal view rather than the total 3D segmentation, to follow aspects of the clinical procedure. PCA was applied to the averaged 2D EAS segmentation, rather than the mid-sagittal plane containing X_{cm} to make the method more robust, and reduce the risk of incorrect rotation due to poor segmentation of the EAS within one mid-sagittal plane. To form the rotation matrix, R, the inverse of the averaged eigenvector, $\vec{v}_{av}^{-1}$ was computed. The rotation matrix, R, was defined as:

$$R = \|R_x\| \, \|R_y\| \, \|R_z\| = \begin{Vmatrix} 1 & 0 & 0 \\ 0 & \vec{v}_{av_{xx}}^{-1} & \vec{v}_{av_{yx}}^{-1} \\ 0 & \vec{v}_{av_{xy}}^{-1} & \vec{v}_{av_{yy}}^{-1} \end{Vmatrix}. \tag{1}$$

where $\vec{v}_{av_{xx}}^{-1}$ and $\vec{v}_{av_{xy}}^{-1}$ define the x and y component respectively of the eigenvector along the length of the anal canal, and $\vec{v}_{av_{yx}}^{-1}$ and $\vec{v}_{av_{yy}}^{-1}$ define the x and y component respectively of the eigenvector along the width of the anal canal. The TPUS and CNN segmentation were rotated in preparation for TUI extraction.

Unfortunately, occasionally the rotation angle determined as above may be too severe, due to a non cylindrical EAS segmentation. Therefore, before TUI extraction occurred an automated quality control process was performed. The ratio between the largest and smallest eigenvector component was calculated, and when the ratio was smaller than a pre-defined threshold, the rotation matrix was set to identity, and the TPUS and segmentation were not rotated. The pre-defined threshold was 2.11 and it was determined in preliminary studies, based on the relationship between the eigenvector component ratio and the rotational acceptability score. In detail, a sample of 10 incorrectly rotated TPUS recordings were used, the mean ratio and standard deviation were calculated, and the threshold was set to the upper bound of the 95% confidence limit.

Identification of Extreme Points. To extract TUI sequences, the extreme points as shown in Fig. 2 were identified. X_{cm} of the rotated CNN segmentation was calculated and the rotated mid-sagittal plane containing X_{cm} was extracted. After rotation the major axes of the EAS were parallel to the coordinate axes, and the first and last coordinate along the y axis of the EAS segmentation were extracted. The total length of the EAS was calculated and divided by 9 to determine the slice separation (i.e. distance between 2D slices in coronal-view of the anal sphincter complex), thus 8 slices were extracted excluding the first and last coordinate position of the EAS. This reduced the risk of selecting a plane too far from the optimal position due to poor segmentation. During examination the spacing between TUI sequences should be larger than 2mm, thus a quality check was performed prior to extraction, and when the slice separation was smaller than 2mm the total length of the EAS was divided by 7 and the TUI sequences included the extremities of the EAS segmentation. If the slice separation was still smaller than 2mm the algorithm outputted the TUI sequences and a notification that the length of the detected EAS may be insufficient or abnormal.

2.3   Data Collection

Analysis of anonymised, archived US images was retrospective, so ethics committee approval was not required by Belgian law. The TPUS recordings were acquired at the pelvic floor clinic at UZ Leuven, Belgium between February and November 2020. The data was separated into training and test sets such that each patient was in one set only. In total 148 3D TPUS recordings were used; 94 for training, 24 for validation and 30 for testing. An expert sonographer with over four years' of experience in US PFD assessment, manually extracted TUI sequences of the anal sphincter for clinical diagnosis using 4D View software (GE Healthcare, Zipf, Austria). The same expert manually segmented the EAS complex with rotations of 30°, using the volume analysis application VOCAL from 4D View Software (GE Healthcare, Zipf, Austria) from the 3D TPUS recordings, these were used as ground truth labels for training.

2.4   Evaluation Methodology

The expert identified the TUI sequences in all TPUS recordings manually via the clinical protocol using 4D View software (GE Healthcare; Zipf, Austria). These TUI sequences are defined as gold standard and were visually compared to the automatically detected TUI sequences. To assess the performance of the proposed method, the expert was asked to rate the overall performance of the automatically detected TUI sequences for each TPUS volume: visually "clinically acceptable" or "unacceptable" for clinical diagnosis. They were also asked to rate the rotation of the anal canal within the mid-sagittal plane and the quality of each TUI slice as either "clinically acceptable" or "unacceptable". The slice rating was dependent on the automatic slice being visually similar to the gold standard and of use for clinical diagnosis (i.e. showing the same pathology if present). The time taken for the automatic pipeline to identify the TUI sequences was compared to the time taken by the expert to manually extract the TUI sequence on a new subset of 19 TPUS recordings acquired within clinic, via the clinical protocol using 4D View software (GE Healthcare; Zipf, Austria). In addition the slice separation of TUI sequences determined automatically and manually were compared.

3   Experiments, Results and Discussion

3.1   Implementation Details

The proposed tool was implemented on a Windows desktop with a 24GB NVIDIA Quadro P6000 ($NVIDIA, California, UnitedStates$). The CNN was implemented using NiftyNet [4], training and inference were ran on the GPU. The CNN architecture was 3D U-Net [14], an Adam optimiser, ReLU activation function, weighted decay factor of 10^{-5}, Dice loss function with a learning rate of 0.0001 and batch size of 2 were used. The data augmentation used were: elastic deformation (deformation sigma = 5, number of control points = 4), random scaling (-20%, $+20\%$) and an implementation of mixup [13]. Validation of the CNN training was performed every 200 epochs and it trained for 6000 epochs. The model from epoch 3200 was used at inference, as the validation loss function was lowest.

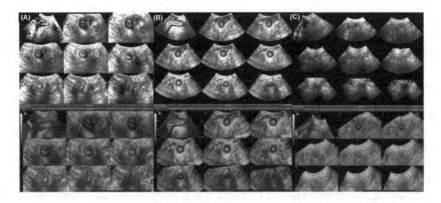

Fig. 3. TUI extraction results, a) the best performing result, b) the average perform-ing result and c) the worst performing result. The corresponding gold-standard TUI sequence for each result are shown in the second row.

Table 1. Overall and rotational clinical acceptability, time taken and slice separation.

Method	Overall clinical Acceptability, %	Rotation clinical Acceptability, %	Time s	Slice separation mm
Automatic method	90.00	93.33	8.64 ± 0.17	2.84 ± 0.50
Manual	100	100	61.00 ± 13.74	2.81 ± 0.37

3.2 Results

Qualitative results of the automatically extracted TUI sequences compared to the gold standard are shown in Fig. 3. Figure 3c shows the worst performing result (based on overall, slice and rotational acceptability), Fig. 3b the average performing result, and Fig. 3a the best performing result. The overall clinical acceptability, the rotation performance, the time taken, and the average slice separation are shown in Table 1. Table 2 shows the clinical acceptability scores of the automatic method for each TUI slice.

3.3 Discussion

The study presents to the-authors-knowledge the first automatic TUI sequence extraction pipeline from a TPUS recording. Qualitatively in Fig. 3 there is min-imal visual difference between the automatically and manually extracted TUI sequences for the average and best performing result and they both show the

Table 2. Slice number and corresponding clinical acceptability.

Slice number	1	2	3	4	5	6	7	8
Clinical acceptability	70.00	86.67	90.00	93.33	96.67	96.67	96.67	90.00

same clinical diagnosis. The worst-performing result was clinically unacceptable for all slices, due to the incorrect rotation and location of the TUI sequence. Incorrect rotation was due a non cylindrical EAS with a ratio of 3.45, which was larger than the pre-defined threshold. Incorrect rotation meant the TUI sequences did not intersect the anal canal perpendicularly, and incorrect location of the first TUI sequence resulted in sequences that were not clinically suitable. The average-performing result had a rotation that was clinically acceptable, however, the position of the final slice was not optimal as it contained part of the IAS unlike the gold standard. This does not impact the overall clinical acceptability as the same diagnosis was made. The best performing result, was rated as clinically acceptable for all slices and for the rotation. The visual difference between the automatic result and the gold standard is negligible and any differences are due to the post-processing of 4D View (GE Healthcare; Zipf, Austria).

The proposed method was 52.36 s faster than the clinical expert, which was significant ($p < 0.001$), and the variance of time taken decreased significantly ($p < 0.001$). The average slice separation of the proposed method was not statistically higher than the manually acquired slice separation ($p = 0.265$). The overall clinical acceptability of the proposed method was 90.00%, on average 7.16 TUI sequences out of 8 were marked as clinically acceptable and the rotation scored a clinical acceptability of 93.33%. 11 TPUS volumes were not rotated as the quality control process detected a ratio of eigenvector components smaller than or equal to 2.11. Slice 1 and 8 describe the extremities of the EAS, and are the most dependent on the segmentation. Table 2 shows slice 1 and 2 were the least clinically accurate, the location of the first slice may improve with a larger training dataset of EAS segmentations. The other slices performed similarly.

The strengths of this work are that it allows a non expert to extract the TUI sequences for diagnosis, and it saves a significant amount of time for all (expert and non-expert) sonographers. Automation may standardise the current procedure and reduce inter-observer variability, this will be studied in future work, on a larger dataset. The main limitation, is the formation of the rotation matrix, as it is dependent on the EAS segmentation. Incorrect segmentation due to artefacts, may lead to a non-cylindrical shape, and the volume may not be rotated at all, or rotated too much. In some patients biologically the EAS may not be cylindrical during contraction, regardless of CNN performance. Thus, in the future, we aim to include segmentation information of the IAS. In addition, to follow clinical guidelines more closely, we aim to ensure the anal canal is not only horizontally aligned in the mid-sagittal plane, but also that it is vertically aligned in the axial plane. This would improve results when the US is acquired sub-optimally (i.e. asymmetric), allowing less-skilled sonographers to perform TUI extraction. As the current method does not correct asymmetric US recordings in the axial plane, the TUI sequences may not intersect the anal canal perpendicularly, leading to sub-optimal TUI sequences. Previous work highlighted that inter observer agreement for sphincteric measurements was fair to excellent for transperineal acquisition [2], however, in future work the evaluation will be expanded to several clinical observers to calculate intra and inter observer variability to reduce bias. Furthermore, the pipeline will be extended to classify anal sphincter tears and disease if present.

4 Conclusion

To conclude, the proposed method achieved an overall clinical acceptability of 90.00%, despite the limitation of the rotation matrix and not rotating the axial plane as performed in clinic to improve asymmetric acquisitions. Thus, we believe with a more detailed pipeline which includes IAS segmentation, the results will outperform this method, and may perform comparable to inter-observer variability. The proposed method was 52.36 s quicker than the clinical expert, which was significant. The proposed method allows non-expert sonographers to perform TUI sequence extraction for anal sphincter tear diagnosis. In future work we will conduct an inter and intra observer variability study, and expand the evaluation dataset to 100 TPUS volumes.

Acknowledgments. We gratefully acknowledge General Electric Healthcare (Zif, Austria), for their continued research support.

References

1. Bonmati, E., et al.: Automatic segmentation method of pelvic floor levator hiatus in ultrasound using a self-normalising neural network. J. Med. Imaging **5**, 12 (2017)
2. Cattani, L., et al.: Exo-anal imaging of the anal sphincter: a comparison between introital and transperineal image acquisition. Int. Urogynecol. J. **31**(6), 1107–1113 (2019). https://doi.org/10.1007/s00192-019-04122-5
3. Dietz, H.P.: Exoanal imaging of the anal sphincters. J. Ultrasound Med. **37**(1), 263–280 (2018)
4. Gibson, E., et al.: NiftyNet: a deep-learning platform for medical imaging. CoRR, abs/1709.03485 (2017)
5. Li, X., Hong, Y., Kong, D., Zhang, X.: Automatic segmentation of levator hiatus from ultrasound images using U-Net with dense connections. Phys. Med. Biol. **64**(7), 075015 (2019)
6. Martinez Franco, E., et al.: Transperineal anal sphincter complex evaluation after obstetric anal sphincter injuries: with or without tomographic ultrasound imaging technique? Eur. J. Obstetr. Gynecol. Reprod. Biol. **257**, 70–75 (2021)
7. Shek, K.L., Zazzera, V.D., Atan, I.K., Rojas, R.G., Langer, S., Dietz, H.P.: The evolution of transperineal ultrasound findings of the external anal sphincter during the first years after childbirth. Int. Urogynecol. J. **27**(12), 1899–1903 (2016). https://doi.org/10.1007/s00192-016-3055-z
8. Sheth, S.: AIUM/IUGA practice parameter for the performance of urogynecological ultrasound examinations: developed in collaboration with the ACR, the AUGS, the AUA, and the SRU. J. Ultrasound Med. **38** (2019)
9. Sindhwani, N., et al.: Semi-automatic outlining of levator hiatus. Ultrasound Obstetr. Gynecol. **48**, 09 (2015)
10. Stuart, A., Ignell, C., Örnö, A.-K.: Comparison of transperineal and endoanal ultrasound in detecting residual obstetric anal sphincter injury. Acta Obstetricia et Gynecologica Scandinavica **98**(12), 1624–1631 (2019)

11. van den Noort, F., et al.: Deep learning enables automatic quantitative assessment of puborectalis muscle and urogenital hiatus in plane of minimal hiatal dimensions. Ultrasound Obstetr. Gynecol. **54**(2), 270–275 (2019)
12. Williams, H., et al.: Automatic C-plane detection in pelvic floor transperineal volumetric ultrasound. In: Hu, Y., et al. (eds.) ASMUS/PIPPI -2020. LNCS, vol. 12437, pp. 136–145. Springer, Cham (2020). https://doi.org/10.1007/978-3-030-60334-2_14
13. Zhang, H., Cisse, M., Dauphin, Y., Lopez-Paz, D.: Mixup: beyond empirical risk minimization, 10 2017
14. Çiçek, Ö., Abdulkadir, A., Lienkamp, S.S., Brox, T., Ronneberger, O.: 3D U-Net: learning dense volumetric segmentation from sparse annotation (2016)

Lung Ultrasound Segmentation and Adaptation Between COVID-19 and Community-Acquired Pneumonia

Harry Mason[1,2]([✉]), Lorenzo Cristoni[3], Andrew Walden[4], Roberto Lazzari[5],
Thomas Pulimood[6,7], Louis Grandjean[8,9],
Claudia A. M. Gandini Wheeler-Kingshott[10,11], Yipeng Hu[1,2],
and Zachary M. C. Baum[1,2]

[1] Centre for Medical Image Computing, University College London, London, UK
harry.mason.18@ucl.ac.uk
[2] Wellcome/EPSRC Centre for Surgical and Interventional Sciences, University College London, London, UK
[3] Frimley Park Hospital, Frimley Health NHS Foundation Trust, Frimley, UK
[4] Royal Berkshire Hospital, Royal Berkshire NHS Foundation Trust, Reading, UK
[5] Hospital de La Santa Creu I Sant Pau, Barcelona, Spain
[6] West Suffolk Hospital, West Suffolk NHS Foundation Trust, Bury St Edmunds, UK
[7] Cambridge University Hospital, University of Cambridge, Cambridge, UK
[8] NHS Foundation Trust, Great Ormond Street Children's Hospital, London, UK
[9] Institute of Child Health, University College London, London, UK
[10] NMR Research Unit, Queen Square MS Centre, UCL Queen Square Institute of Neurology, London, UK
[11] Department of Brain and Behavioural Sciences, University of Pavia, Pavia, Italy

Abstract. Lung ultrasound imaging has been shown effective in detecting typical patterns for interstitial pneumonia, as a point-of-care tool for both patients with COVID-19 and other community-acquired pneumonia (CAP). In this work, we focus on the hyperechoic B-line segmentation task. Using deep neural networks, we automatically outline the regions that are indicative of pathology-sensitive artifacts and their associated sonographic patterns. With a real-world data-scarce scenario, we investigate approaches to utilize both COVID-19 and CAP lung ultrasound data to train the networks; comparing fine-tuning and unsupervised domain adaptation. Segmenting either type of lung condition at inference may support a range of clinical applications during evolving epidemic stages, but also demonstrates value in resource-constrained clinical scenarios. Adapting real clinical data acquired from COVID-19 patients to those from CAP patients significantly improved Dice scores from 0.60 to 0.87 ($p < 0.001$) and from 0.43 to 0.71 ($p < 0.001$), on independent COVID-19 and CAP test cases, respectively. It is of practical value that the improvement was demonstrated with only a small amount of data in both training and adaptation data sets, a common constraint for deploying machine learning models in clinical practice. Interestingly, we also report that the inverse adaptation, from labelled CAP data to unlabeled COVID-19 data, did not demonstrate an improvement when tested on either condition. Furthermore, we offer a possible explanation that correlates the segmentation performance to

J. A. Noble et al. (Eds.): ASMUS 2021, LNCS 12967, pp. 45–53, 2021.
https://doi.org/10.1007/978-3-030-87583-1_5

label consistency and data domain diversity in this point-of-care lung ultrasound application.

Keywords: Deep-learning · Segmentation · Domain adaptation · Lung ultrasound · COVID-19 · Pneumonia

1 Introduction

Over the past decade, the use of point of care ultrasound (POCUS) has increased alongside the growing evidence relating its use to improved patient outcomes. The publication of the BLUE protocol displayed the efficiency of POCUS in the diagnosis of the 5 most common lung pathologies compared to chest auscultation and chest x-ray, achieving an accuracy of 90.5% [1]. POCUS was shown to be useful in the triaging of patients with suspected COVID-19 by following the BLUE protocol [2, 3]. Both COVID-19 and CAP present multiple B-lines in the early stages and areas of consolidation appear as infection progresses. Although computerized tomography (CT) scans have shown sensitivity of up to 97% [4] for the diagnosis of COVID-19, it can be impractical for use in 'front-line' settings, as it requires patients to be moved throughout the hospital, may risk precautious patients desaturating in scanner, and is time-consuming. Conversely, the BLUE protocol can be performed in a few minutes at the patient's bedside, making POCUS advantageous for use during a pandemic when resources are low and infection risk is high.

Several studies have investigated the use of deep learning to assist in triage, diagnosis, grading and monitoring of COVID-19 patients [5–10]. Methods include classifying and stratifying COVID-19 patients, or localizing pathological image features, all based on lung ultrasound (LUS) data from healthy subjects or other respiratory diseases, such as pulmonary edema and community-acquired pneumonia (CAP). To improve the specificity of computer-assisted tools, aggregating approaches combining pixel-, frame-, zone-, and patient-level severity scores have been proposed [11]. Localization, and therefore, segmentation, of pathology-sensitive LUS features then, on the pixel level, is fundamental. Moreover, the intuitive representation of segmentation, such as those of B-lines used in this study, may provide a visually interpretable solution in the form of a prediction for the clinician. Such segmentations may not only help the confidence the clinician would place on the automated computer prediction by localizing it, but also provides a feedback opportunity to further develop the assistive algorithm for improved sensitivity and specificity.

Most existing research in machine learning, such as the work we present here, requires retrospectively labelled data. However, deploying such algorithms in real-world clinical use has direct challenges. Most prominently, efficiently obtaining high-quality labelled data [12]. For example, at the beginning of an epidemic, or during its fast-changing stage, representative imaging data from positive patients is usually scarce. Furthermore, obtaining expert labels may be even more costly during the peak of an outbreak. In scenarios such as these, the ability to use a pre-trained model, or an existing data set, perhaps from a relevant condition (CAP, in this work), could substantially reduce the requirements for necessary data and labeling from the target application, using fine-tuning or unsupervised domain adaptation. A different type of scenario, also investigated in this work, may be that data from a previous or ongoing epidemic (COVID-19, in this work) are available for training the pre-trained models or being used as the source domain data to be adapted to a different type of condition that has less, limited, or unlabeled data. Examples include pneumonia caused by a new epidemic, an additional variant to the existing one, or other types of pneumonia in an area that lacks access to other data sources or labeling expertise. In this study, we test the transfer learning and domain adaptation abilities to and from the COVID-19 patient data, with the CAP patient data as an example of the other LUS data.

2   Methods

We consider two strategies for training convolutional neural networks to segment B-lines from LUS images. The first strategy uses a supervised learning approach, requiring manual labeling of all input data, to segment the B-lines in the LUS images. The second strategy uses an unsupervised domain adaptation to adapt a segmentation network to an unlabeled target domain, requiring labels for only the source segmentation domain in training.

2.1   Supervised Segmentation with U-Net

A commonly used neural network for image segmentation, U-Net [13], was trained to automatically segment B-lines in COVID-19 and CAP LUS images. At inference, the network then predicts whether a given pixel in the image may be classified as part of a B-line. The use of well-established network architectures, such as U-Net, allows this work to focus primarily on investigating the feasibility of automatic segmentation of these regions of pathological interest.

2.2   Unsupervised Segmentation via Image and Feature Alignment

Synergistic image and feature alignment (SIFA) [14] has been used for domain adaptation tasks to guide the adversarial learning of an end-to-end framework for unsupervised image segmentation. SIFA reduces domain shift by using a generative adversarial network to synthetically translate images from a source domain to the target domain. The network is composed of a generator, which learns to translate the source domain image into a corresponding image of the target domain, an encoder that learns a shared feature-space, a decoder that learns the reverse-translation from target to source, and a segmenter that performs pixel-wise classification to identify different labels in the images. Additionally, three discriminator networks differentiate between the target and source inputs to the encoder, and the outputs of the decoder and segmenter. SIFA is trained to automatically segment B-lines in COVID-19 and CAP LUS images. However, in training, labels for only the source domain are required to learn the segmentation of the target domain.

2.3 Implementation Details

All neural networks were implemented in TensorFlow [15] and Keras [16]. Reference-quality open-source code was adopted where possible for reproducibility.

Our implementation of U-Net contained 4 layers of convolutional blocks, with an increasing number of channels of 16, 32, 64, and 128. Each convolutional block used Batch Normalization across the channel axis between convolutional layers and a Dropout of 0.5 following each Batch Normalization. We employed data augmentation using rotation, shifting, and scaling to reduce over-fitting. All U-Net models were trained for 250 epochs with a mini-batch size of 16, using an equal-weight binary cross-entropy and Dice loss and the Adam optimizer [17] with a learning rate of 0.005.

Our implementation of SIFA and hyperparameters described below are consistent with the original default implementation and hyperparameters, as described in [14]. As in the original implementation, we employed data augmentation using rotation, shifting, and scaling to reduce over-fitting. All SIFA models were trained for 10,000 epochs, with a mini-batch size of 12. The generator, encoder, and decoder were trained using a weighted cycle-consistency and adversarial loss with the Adam optimizer at a learning rate of 0.0002. The segmenter was trained using an equal-weight cross-entropy and Dice loss with the Adam optimizer at a learning rate of 0.001.

2.4 Data

The US images were acquired from two hospitals by two clinicians, using a Butterfly iQ US probe (Butterfly Inc., Guilford, CT, USA). Experiments were conducted using images from six COVID-19 positive patients and seven patients with CAP. Due to the low prevalence of B-lines within patient scans only images with B-lines were used for training, to evaluate the segmentation algorithms. The resulting datasets contained 977 and 326 images for COVID-19 and CAP, respectively. All COVID-19 diagnoses were confirmed by PCR tests.

Ground-truth B-line segmentations were manually labeled by a medical student familiar with LUS. Segmentations were reviewed and verified by experienced US imaging researchers with over five years of experience with clinical US imaging. Example images and segmentations are provided in Fig. 1.

2.5 Experiments

Given the limited data availability, we adopt a two-way split for training and test sets, without a validation set. This prevents fine-tuning of network parameters or other hyperparameters to optimize performance, to avoid information leakage and provide fair estimates of model performances. As such, cross-validation was performed to assess the performance of models for the segmentation of B-lines in patients with COVID-19 and CAP under different supervision conditions in training. The COVID-19 and CAP datasets were split into three cross-validation sets, on a patient-level. Splitting the data in this way ensures that no patient images are found amongst the different dataset splits. Efforts were made to ensure that each of the three COVID-19 and CAP datasets were of approximately the same size. The COVID-19 datasets each consisted of 2 patients, with

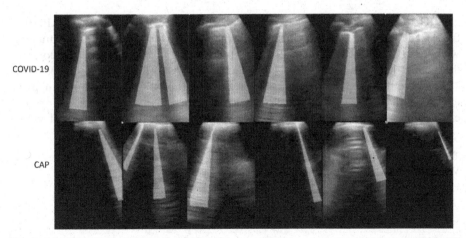

Fig. 1. Multiple sample LUS images and their corresponding manual segmentations. COVID-19 images and segmentations are shown on the top row, CAP images and segmentations are shown on the bottom row.

319, 319, and 339 images, respectively. The CAP datasets consisted of one, two, and three patient(s), with 148, 111, and 67 images, respectively. In total, seven experiments are presented to evaluate and assess the performance of U-Net and SIFA for segmentation of B-lines in patients with COVID-19 and CAP.

Four of these seven experiments are performed with U-Net. First, we train U-Net with COVID-19 images. Second, we train U-Net with CAP images. Third, we train U-Net with COVID-19 images and fine-tune with CAP images. Finally, we train U-Net with CAP images and fine-tune with COVID-19 images. In both instances, fine-tuning took place over 50 epochs at a learning rate of 0.0005. Corresponding COVID-19 and CAP datasets are used in training and fine-tuning when applicable.

Three of these seven experiments are performed with SIFA. First, we train SIFA with a source domain of COVID-19 images and a target domain of CAP images. To assess if the discrepancy in dataset sizes affects the training of SIFA, we then train SIFA with a source domain of COVID-19 images, where we use only a reduced subset of the COVID-19 images and a target domain of CAP images. Finally, we train SIFA with a source domain of CAP images and a target domain of COVID-19 images. To evaluate each of the previously described methods, segmentations were evaluated based on a binary Dice score. We additionally report sensitivity, specificity and p-values from statistical t-tests at a significance level of 0.05, when comparison was made.

3   Results

Table 1 summarizes the Dice scores from the cross-validation experiments across all methods. Training with SIFA (COVID-19 Source/CAP Target) provided significantly higher Dice scores on COVID-19 and CAP test data than all four U-Net methods ($p < 0.001$) and with SIFA (CAP Source/COVID-19 Target) ($p < 0.001$). Training with SIFA (CAP Source/COVID-19 Target) provided significantly lower Dice scores on the

COVID-19 and CAP test data than U-Net (COVID-19) ($p < 0.001$), U-Net (CAP) ($p < 0.001$), and U-Net (COVID-19 Fine-Tune w/CAP), ($p < 0.001$). Additionally, training with SIFA (CAP Source/COVID-19 Target) provided significantly lower Dice scores on the COVID-19 test data than U-Net (CAP Fine-Tune w/COVID-19), ($p < 0.001$), but no significant difference was found to U-Net (CAP Fine-Tune w/ COVID-19) when applied to CAP test data ($p = 0.23$). Table 2 summarizes the sensitivity and specificity from the cross-validation experiments and is consistent with the observation summarized above.

Table 1. Summary of segmentation cross-validation Dice scores. STD: Standard Deviation. Values are presented as Mean ± STD.

Method	COVID-19	CAP	
	Dice	Dice	
U-Net (COVID-19)	0.60 ± 0.26	0.36 ± 0.24	
U-Net (CAP Fine-Tune w/COVID-19)	0.56 ± 0.25	0.35 ± 0.26	
U-Net (CAP)	0.52 ± 0.17	0.43 ± 0.27	
U-Net (COVID-19 Fine-Tune w/CAP)	0.55 ± 0.15	0.45 ± 0.24	
SIFA (COVID-19 Source/CAP Target)	0.87 ± 0.13	0.71 ± 0.22	
SIFA (Reduced COVID-19 Source/CAP Target)	0.83 ± 0.15	0.72 ± 0.21	
SIFA (CAP Source/COVID-19 Target)	0.32 ± 0.21	0.33 ± 0.17	

Table 2. Summary of segmentation cross-validation sensitivity (sens.) and specificity (spec.).

Method	COVID-19		CAP	
	Sens	Spec	Sens	Spec
U-Net (COVID-19)	0.55	0.93	0.38	0.92
U-Net (CAP Fine-Tune w/COVID-19)	0.47	0.97	0.44	0.95
U-Net (CAP)	0.50	0.90	0.50	0.93
U-Net (COVID-19 Fine-Tune w/CAP)	0.48	0.91	0.40	0.95
SIFA (COVID-19 Source/CAP Target)	0.86	0.98	0.78	0.96
SIFA (Reduced COVID-19 Source/CAP Target)	0.84	0.95	0.75	0.95
SIFA (CAP Source/COVID-19 Target)	0.30	0.95	0.31	0.94

Additionally, we present qualitative examples of segmentations from both methods, trained with all the different aforementioned approaches, on COVID-19 images and CAP images in Figs. 2 and 3, respectively. These visualizations demonstrate the ability of these networks to delineate B-lines, suggesting that, in some instances, they may be effectively used for assisting in the interpretation of LUS in clinical practice.

As a retrospective analysis, we aim to explain the observed difference in improvement (or lack of it) between the two directions of adaptation when testing the resulting models

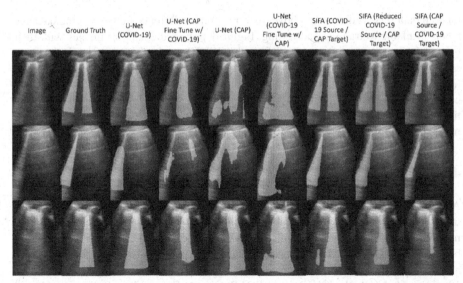

Fig. 2. Three example LUS images, each illustrating segmentations from each of the different methods on COVID-19 images. Each image shows the original LUS image and the segmentation output corresponding to the ground truth, or the method used. Each column presents the image and segmentation for the method listed above them.

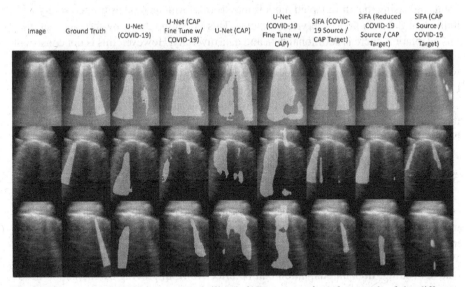

Fig. 3. Three example LUS images, each illustrating segmentations from each of the different methods on CAP images. Each image shows the original LUS image and the segmentation output corresponding to the ground truth, or the method used. Each column presents the image and segmentation for the method listed above them.

on both data sets, as described above, in terms of the difference in the imaging data and labels available to training and testing. Figure 1 provided examples images with their ground-truth segmentations overlaid, from the COVID-19 and CAP data sets, in the upper and lower rows, respectively. It is visibly evident that CAP data posses substantially higher variability in location, size of the identified B-line patterns and their background context. This is consistent with all the data used in our study. This is also consistent with the annotators' experience indicating that labelling on the CAP data set is considered a more challenging task than that on the COVID-19 data set.

4 Discussion

Additionally, during initial experimentation, we evaluated the performance of a joint-training strategy for supervised segmentation with U-Net in addition to the fine-tuning methods. Here, instead of fine-tuning on a pre-trained model, we train the model on both the COVID-19 and CAP datasets simultaneously. Notably, the performance over all cross-validation folds resulted in comparable Dice scores to training only on COVID-19 when tested on both COVID-19 and CAP test sets. For brevity, we did not include a full validation of this training strategy in our above-presented experiments.

One of the interesting findings in this work is that the substantial difference between the adaptation from two opposite directions, from CAP to COVID-19 and from COVID-19 to CAP, with only the latter showing benefit of adaptation on both test datasets. It is not unsurprising that the adapted models may outperform the models trained solely with individual datasets in a supervised manner. This is compounded by the fact that with this adaptation, there is additional data and data diversity. However, this is not consistent with the performance observed when adapting from CAP to COVID-19. Intuitively, we may associate this with the label uncertainty and variability observed within the CAP images, as previously discussed in Sect. 3. It is important to note that, especially constrained by small data sets, the efficacy of domain adaptation is highly dependent on the data diversity and label uncertainty, one needs to be further understood and validated before being deployed in clinical applications.

5 Conclusion

In this work, we have presented the development and validation experiments for segmenting real clinical LUS data, acquired from both CAP and COVID-19 patients, and in particular the approaches for combining the two for training deep neural networks. We report a set of interesting experimental results that demonstrated that, in a small data set setting, domain adaptation can be effective in improving segmentation accuracy by incorporating additional unlabelled data. However, compared to the direction of the desirable adapting, the availability of diverse data and high-quality, consistent and representative labels were more strongly correlated with such improvement. The experimental results provided preliminary evidence for the feasibility and practicality of aggregating different types of data in this POCUS application.

Acknowledgments. This work is supported by the Wellcome/EPSRC Centre for Interventional and Surgical Sciences (203145Z/16/Z). C.A.M. Gandini Wheeler-Kingshott is supported by the MS Society (#77), Wings for Life (#169111), Horizon2020 (CDS-QUAMRI, #634541), BRC (#BRC704/CAP/CGW), and allocation from the UCL QR Global Challenges Research Fund (GCRF). Z.M.C. Baum is supported by the Natural Sciences and Engineering Research Council of Canada Postgraduate Scholarships-Doctoral Program, and the University College London Overseas and Graduate Research Scholarships.

References

1. Lichtenstein, D., Goldstein, I., Mourgeon, E., Cluzel, P., Grenier, P., Rouby, J.J.: Comparative diagnostic performances of auscultation, chest radiography, and lung ultrasonography in acute respiratory distress syndrome. J. Am. Soc. Anesthesiol. **100**(1), 9–15 (2004)

2. Antúnez-Montes, O.Y., Buonsenso, D.: Routine use of point-of-care lung ultrasound during the COVID-19 pandemic. Medicina Intensiva (2020)

3. Jackson, K., Butler, R., Aujayeb, A.: Lung ultrasound in the COVID-19 pandemic. Postgrad. Med. J. **97**(1143), 34–39 (2021)

4. Ai, T., et al.: Correlation of chest CT and RT-PCR testing for coronavirus disease 2019 (COVID-19) in China: a report of 1014 cases. Radiology **296**(2), E32–E40 (2020)

5. Born, J., et al.: POCOVID-Net: automatic detection of COVID-19 from a new lung ultrasound imaging dataset (POCUS). arXiv preprint arXiv:2004.12084 (2020)

6. Baum, Z., et al.: Image quality assessment for closed-loop computer-assisted lung ultrasound. In: SPIE Medical Imaging 2021: Image-Guided Procedures, Robotic Interventions, and Modeling, 115980R (2021)

7. Roy, S., et al.: Deep learning for classification and localization of COVID-19 markers in point-of-care lung ultrasound. IEEE Trans. Med. Imaging **39**(8), 2676–2687 (2020)

8. Arntfield, R., et al.: Development of a convolutional neural network to differentiate among the etiology of similar appearing pathological B lines on lung ultrasound: a deep learning study. BMJ Open **11**(3), e045120 (2021)

9. Horry, M.J., et al.: COVID-19 detection through transfer learning using multimodal imaging data. IEEE Access **8**, 149808–149824 (2020)

10. Bagon, S., et al.: Assessment of COVID-19 in lung ultrasound by combining anatomy and sonographic artifacts using deep learning. J Acoust. Soc. Am. **148**(4), 2736 (2020)

11. Xue, W., et al.: Modality alignment contrastive learning for severity assessment of COVID-19 from lung ultrasound and clinical information. Med. Image Anal. **69**, 101975 (2021)

12. Hu, Y., Jacob, J., Parker, G.J., Hawkes, D.J., Hurst, J.R., Stoyanov, D.: The challenges of deploying artificial intelligence models in a rapidly evolving pandemic. Nat. Mach. Intell. **2**, 298–300 (2020)

13. Ronneberger, O., Fischer, P., Brox, T.: U-Net: convolutional networks for biomedical image segmentation. Med. Image Comput. Comput.-Assist. Interv. **2015**, 234–241 (2015)

14. Chen, C., Dou, Q., Chen, H., Qin, J., Ann Heng, P.: Unsupervised bidirectional cross-modality adaptation via deeply synergistic image and feature alignment for medical image segmentation. arXiv preprint arXiv:2002.02255 (2020)

15. Abadi, M., et al.: TensorFlow: Large-scale machine learning on heterogeneous systems (2015). https://tensorflow.org

16. Chollet, F.: Keras (2015). https://keras.io

17. Kingma, D.P., Ba, J.: Adam: a method for stochastic optimization. In: International Conference for Learning Representations (2015)

An Efficient Tracker for Thyroid Nodule Detection and Tracking During Ultrasound Screening

Ting Liu, Xing An, Bin Lin, Yanbo Liu, Wenlong Xu, Yuxi Liu, Longfei Cong, and Lei Zhu$^{(\boxtimes)}$

Shenzhen Mindray BioMedical Electronics, Co., Ltd., Shenzhen, China
zhulei@mindray.com

Abstract. Thyroid tumor is a common disease in clinic. Junior doctors could easily miss or get false detection due to the unclear boundary and similarity between nodules and tissues during thyroid screening. In this paper, we propose an efficient tracker for simultaneously detecting and tracking nodules to assist doctors in examination and improve their work efficiency. An attention based fusion block which adaptively combines the features of previous and current frames is introduced to acquire better detection and tracking result. To increase the detection accuracy, we propose an advanced post-processing strategy instead of using general post-processing methods to train the network to obtain the best prediction. Moreover, a minibatch self-supervised learning module is embedded to reduce the false positive rate (FPR) by strengthening the ability of distinguishing nodules from similar tissues. The proposed framework is validated on a dataset of 1555 thyroid ultrasound movies with 13314 frames. The result of 91% recall with 3.8% FPR running at 30 fps demonstrates the effectiveness of our method.

Keywords: Detection · Tracking · Thyroid ultrasound image · Self-supervised learning

1 Introduction

Thyroid nodules are very common in clinic with the incidence rising rapidly throughout the world. In 2020, 586,202 patients suffered from thyroid cancer, accounting for 2.9% of all cancers [1]. With the growth of health awareness and the widely use of advanced ultrasound equipment, the spotting of thyroid nodules has increased, which brings a great challenge for doctors. Moreover, reviewing large amounts of low-resolution videos is time-consuming, radiologists could lose their concentration which may impact the objectivity on diagnosis. Computer-aided diagnosis can alleviate the workload of doctors and improve the efficiency of their work [2]. Therefore, the development of an automatic and accurate analysis method is necessary for thyroid ultrasound screening.

In recent years, with the development of deep learning, researchers have been investigating convolutional neural networks on thyroid ultrasound image analysis, such as nodule classification and detection. For the task of thyroid image classification, Chi *et al.*

J. A. Noble et al. (Eds.): ASMUS 2021, LNCS 12967, pp. 54–62, 2021.
https://doi.org/10.1007/978-3-030-87583-1_6

[2] fine-tuned a GoogLeNet to extract features of ROIs and predicted the malignancy using the Cost-Sensitive Random Forest algorithm. Ma *et al.* [3] used two pre-trained convolutional neural networks to fuse low-level and high-level features for the classification of thyroid nodules. For the challenge of nodule detection, Abdolali *et al.* [4] enhanced the reliability of detection on a small dataset by modifying Mask R-CNN architecture and combining it with transfer learning. Li *et al.* [5] developed a detector based on Faster R-CNN, and adopted strategies such as layer concatenation and spatial constraint to reach a higher accuracy. Xie *et al.* [6] proposed an SSD based neural network with redesigned loss function and post-processing method to improve the detection recall rate. Wang *et al.* [7] presented an artificial intelligence diagnosis system based on the YOLOv2 to locate and classify nodules simultaneously. Generally, typical techniques widely used in thyroid nodule detection require post-processing method such as non-maximum suppression (NMS) to obtain the optimal prediction in inference.

Although many researches have been done in thyroid ultrasound, the following problems still remain. (i) The majority of methods only focuses on detection in images, ignoring the relationship between previous and current frames in movies. (ii) The performance of NMS used in most framework is limited due to the fixed rules set in advance. Nested predictions of thyroid nodules can still remain after NMS as shown in the first row of Fig. 1, where the tissue inside or outside a nodule seems like another lesion. (iii) Many normal tissues can be recognized as nodules due to the similarity between them as shown in the second row of Fig. 1, but less attention is paid to differentiate them.

In view of above issues, we propose an efficient tracker for nodule detection and tracking during thyroid ultrasound scanning. Firstly, we introduce an attention based fusion block to adaptively combine the features of previous and current images to get better detection and tracking result. Secondly, an advanced post-processing strategy that trains the model rather than uses NMS method to find the optimal result is proposed to improve detection accuracy. Finally, a minibatch self-supervised learning module is embedded as a branch in training period to enhance the ability of discriminating nodules from similar tissues, thus to reduce the false positive rate (FPR).

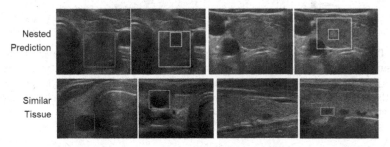

Fig. 1. Illustration of existing problems. Red: the ground truth. Yellow: nested prediction. Green: normal tissues that are similar to nodules. (Color figure online)

2   Methodology

2.1   Overall Architecture

The proposed network is primarily based on CenterTrack [8] which sets a new state of the art on both MOT17 and KITTI datasets. Our method is illustrated in Fig. 2. The network takes the current frame, the previous frame and the heatmap [8] generated from objects in the prior frame as input, and outputs the predicted rectangles, the classification probability of each prediction and the center offset of tracked boxes in adjacent frames. The inputs are combined by a fusion module before being fed into the backbone network which is an encoder-decoder structure, and we adopt ResDCN-18 [8] as the backbone in this work. The outputs are divided into two parts, 1) the predicted boxes and classification probabilities to generate the detection results, and 2) the tracking offsets to determine whether the nodules on the prior and current frame are the same one.

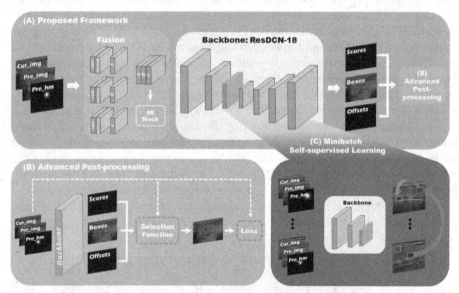

Fig. 2. Illustration of our method. (A) The proposed framework. (B) The advanced post-processing module. (C) The minibatch self-supervised learning module. 'Cur_img', 'Pre_img' and 'Pre_hm' are abbreviations of current image, previous image and heatmap. 'Boxes', 'Scores' and 'Offsets' represent the predicted boxes, the classification probability of each prediction and the offset of the prediction on the current image from the previous frame.

2.2   Fusion Module

The fusion module includes convolution layers and max-pooling layers. It extracts the feature of each input separately and generates three feature maps with size of 1/4 of the input. CenterTrack fuses the feature maps by adding them together and makes each map contribute the same to the next stage. However, we argue that the prior image and heatmap are auxiliary inputs for guiding the model to detect nodules in the current frame. The goal of comprising them is to enhance the performance of detection and tracking. Hence, the feature of current frame should be more important than the other two. Therefore, we concatenate the three feature maps and utilize a squeeze-and-excitation (SE) block to adjust the importance of each channel adaptively. The parameter of reduction in SE block is set to 16 as [9].

2.3 Advanced Post-processing Module

In the detection and tracking task, the optimal prediction should not only consider the classification possibility but also the intersection over union (IoU) and tracking off-set with the ground-truth. Inspired by OneNet [10], we introduce an advanced post-processing module to find the optimal result considering of all three aspects above before calculating the loss, thus the model can be trained to acquire the best prediction directly. Meanwhile, no NMS method is required in inference, and end-to-end strategy is achieved. As shown in Fig. 2 (B), the advanced post-processing module takes predictions from the backbone as input. Before calculating the loss, the selection function calculates a score of each prediction, and only the one with smallest score is considered as the correct prediction and the others are assumed as wrong predictions when computing the loss. The selection function is defined as:

$$S = \lambda_{cls} * S_{cls} + \lambda_{L1} * S_{L1} + \lambda_{giou} * S_{giou} + \lambda_{track} * S_{track} \tag{1}$$

where S_{cls} is the focal loss [11] of predicted classifications and ground truth category labels, S_{L1} and S_{giou} are the L1 loss and the GIoU [12] loss between normalized predictions and ground truth boxes, respectively. S_{track} is the L1 loss between predicted tracking offsets and the real displacement of tracked objects. λ_{cls}, λ_{L1}, λ_{giou}, and λ_{track} are coefficients of each component. Following [8, 10], λ_{cls}, λ_{L1}, λ_{giou} and λ_{track} are set to 2, 5, 2 and 2, separately.

When calculating detection and tracking loss, each ground truth box only corresponds to one predicted result. The loss function is similar to the selection function, and defined as:

$$L_{dt} = \mu_{cls} * L_{cls} + \mu_{L1} * L_{L1} + \mu_{giou} * L_{giou} + \mu_{track} * L_{track} \tag{2}$$

where L_{cls}, L_{L1}, L_{giou}, and L_{track} have the same definition as S_{cls}, S_{L1}, S_{giou} and S_{track}, separately. And the coefficients μ_{cls}, μ_{L1}, μ_{giou} and μ_{track} are equal to λ_{cls}, λ_{L1}, λ_{giou}, and λ_{track}, separately.

2.4 Minibatch Self-supervised Learning Module

To reduce false positive rate, we embed a minibatch self-supervised learning module as a branch in the training period to differentiate nodules from similar tissues. As shown in Fig. 2 (C), the features of the positive (nodules) and negative (normal tissues) samples are obtained in the down-sampling stage. Inspired by [13], a batch-based similarity loss function is proposed to make the Euclidean distance of positive features closer to the most dissimilar positive features and further away from the most similar negative features in a batch. Besides, a constant is set as threshold to ignore tissues that are dissimilar to nodules in nature. The batch-based similarity loss function is defined as:

$$L_{bs} = \frac{1}{2}\|f(p_i), S_p(p_i)\|^2 + \frac{1}{2}[max(0, margin - \|f(p_i), S_n(p_i)\|)]^2 \qquad (3)$$

where $f()$ denotes the global average pooling. $f(p_i)$ is the feature vector of the ith nodule in a batch. $S_p(x)$ and $S_n(x)$ obtains the most dissimilar positive and the most similar negative feature vector to the $f(x)$ in a batch, separately. To be specific, $S_p(p_i)$ is a positive feature vector whose Euclidean distance is the farthest to $f(p_i)$ in the batch and $S_n(p_i)$ is a negative feature vector that has the nearest Euclidean distance to $f(p_i)$ in the batch. $\|\ \|$ represents the Euclidean distance. Margin sets the distance threshold between $f(p_i)$ and $S_n(p_i)$. The loss function will ignore the samples whose Euclidean distance of $f(p_i)$ and $S_n(p_i)$ is larger than the margin as it indicates that the positive sample and the selected negative sample are not similar. Following [13] margin is set to 10.

Furthermore, we randomly produce false samples on images instead of using manual annotation. For a training batch, if the portion of a nodule on an image is less than α, a negative sample that is not intersected with the nodule will be randomly generated, with the size of β times of the nodule. We set $\alpha = 0.3$ because the ratio of false positives is small and less than 1/3 of an image in our dataset, and $\beta \in [0.9, 1.1]$ to maintain the balance between positive and negative samples.

The total loss is summarized as:

$$\mathcal{L} = L_{dt} + \gamma * L_{bs} \qquad (4)$$

where L_{dt} is the detection and tracking loss and L_{bs} denotes the minibatch self-supervised learning loss. γ is a factor and set to 0.2, following [8, 14].

3 Experiments

Dataset. We validated the proposed method on 1555 thyroid ultrasound movies from 1555 patients collected via Mindray Resona 7. There are totally 13,314 frames that extracted from the movies at a fixed (3, 4 or 5) interval. All the images were annotated by five doctors firstly and the final annotations were reviewed by an experienced doctor. We calculated the size of nodules (varied from 462 pixels to 232,672 pixels, mean: 29,979 pixels) and classified the movies into 3 categories according to the tri-sectional quantile of size: small (<1503 pixels), middle (from 1503 pixels to 6923 pixels), and large (>6923 pixels). We randomly split the movies into 80%, 10%, and 10% for training (1244 movies, 10597 images), validation (155 movies, 1363 images) and testing (156 movies, 1354 images) following the stratified sampling.

Implementation Details. We cropped all data and only the ultrasound image contents were remained. All images were resized to 544 × 640 according to the mean image size of our data. Data augmentations were applied including random horizontal flipping, shifting, scaling, and brightness and contract transformation. Adam optimizer with a learning rate of 5e−5 was utilized for training. We trained the network for 300 epochs with batch size of 32 and chose the model with highest AP50 on validation set to do testing. All experiments were performed on a 12 GB NVIDIA TITAN V GPU.

Quantitative and Qualitative Analysis. We measured the FPR (numbers of false positives divided by numbers of images), recall and precision (Prec) to evaluate the detection performance, and recall in tracking (RCLL), mostly tracked (MT) and mostly lost (ML) for tracking performance [15]. The IoU score is set to 0.3 when calculating recall and precision. The novel CenterNet [16], Centertrack, and YoloSiam [17] were re-implemented and evaluated on the same dataset for comparison. Results are shown in Table 1. Our method outperforms all these algorithms not only in above metrics but also in inference speed.

As shown in Fig. 3, our method works well even if the nodule has an undefined margin or is very tiny that less experienced doctors might neglect. Moreover, our method can continuously track on the nodules appear through the whole films. Additionally, our method also works well on multi-target.

Table 1. Performance comparison on the thyroid dataset (%).

Method	FPR	Recall	Prec	AP	AP50	RCLL↑	MT↑	ML↓	Speed
CenterNet	16.1	79.5	83.7	45.2	74.8	NA	NA	NA	28 fps
YoloSiam	21.6	76.3	78.2	41.7	70.5	66.8	61.0	25.0	16 fps
CenterTrack	13.5	82.1	85.3	48.1	77.4	76.6	65.4	16.0	28 fps
Ours	**3.8**	**91.2**	**95.8**	**56.8**	**88.2**	**86.4**	**76.9**	**5.8**	**30 fps**

Ablation Study. An ablation study is conducted on pretrained ResDCN-18 to compare our approach with a representative baseline method. The quantitative results are shown in Table 2. The improvement of the fusion module (F) obtains a better FPR (12.9%) and recall (83.3%) with the same backbone compare to baseline. After using the advanced post-processing module (AP), the FPR is reduced to 8.4% and recall is 2.1% higher than before. The further improvements of FPR from 8.4% to 3.8% and recall from 85.4% to 91.2% indicate the minibatch self-supervised learning module (MSL) has strengthened the ability of distinguishing nodules from similar tissues. We experiment on another backbone (DLA-34) to validate the effectiveness of the proposal. The results also demonstrate the advantage of our method (Table 3).

We calculate the mean channel weight of each input given by the SE block in fusion module. The current frame carries the highest weight of 0.52, the previous frame holds the middle (0.49), and the previous heatmap scores the smallest one at 0.46, which validates our hypothesis. Moreover, the first two columns in Fig. 4 illustrate the proposed method can effectively solve the problem of nested predictions that common scheme cannot handle. The last three columns in Fig. 4 present cases that the baseline detects

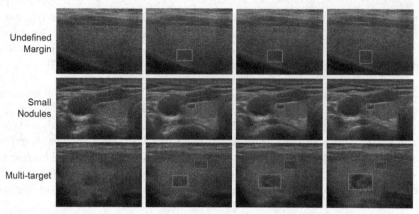

Fig. 3. Detection and tracking results. The first column is an example frame, and the following three columns refer to the second, the forth, and the sixth frame after it.

Table 2. Results of ablation studies based on pretrained ResDCN-18 (%).

Method	FPR	Recall	Prec	AP	AP50	RCLL↑	MT↑	ML↓
Baseline	13.5	82.1	85.3	48.1	77.4	76.6	65.4	16.0
+F	12.9	83.3	85.8	48.3	78.2	78.2	67.9	14.7
+F+AP	8.4	85.4	90.5	51.0	82.5	80.9	71.2	10.3
+F+AP+MSL	**3.8**	**91.2**	**95.8**	**56.8**	**88.2**	**86.4**	**76.9**	**5.8**

Table 3. Results of ablation studies based on DLA-34 (%).

Method	FPR	Recall	Prec	AP	AP50	RCLL↑	MT↑	ML↓
Baseline	14.1	81.3	85.1	46.6	75.7	76.1	63.5	17.9
+F	12.6	82.7	86.0	48.3	78.0	77.8	67.3	15.4
+F+AP	9.8	84.9	88.9	50.7	81.2	79.8	69.2	11.5
+F+AP+MSL	**4.9**	**88.5**	**94.6**	**55.2**	**87.5**	**83.8**	**72.4**	**7.7**

normal tissues as nodules but ours not, which demonstrates our methods can differentiate nodules from the similar tissues better. Meanwhile, Fig. 5 compares the proportion of false positives in testing set, which can show the improvement of our method in FPR clearly.

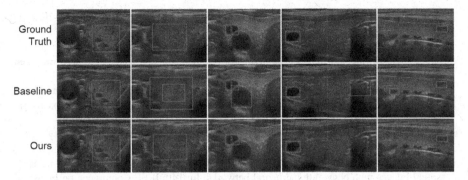

Fig. 4. Results comparison of baseline and ours.

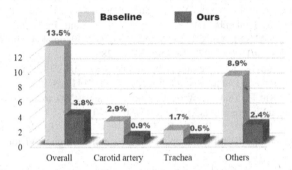

Fig. 5. False positive comparison.

4   Conclusion

To conclude, we propose an efficient tracker to detect and track nodules simultaneously during thyroid ultrasound screening. The attention based fusion block adaptively combines features of the previous and current frames, thus better detection and tracking result is acquired. Moreover, the advanced post-processing mechanism that trains the model instead of using NMS method to select the optimal prediction successfully boosts the detection accuracy. Additionally, the minibatch self-supervised learning module effectively reduces the FPR by enhancing the ability of distinguishing nodules from similar tissues. The result of fast speed, high accuracy, and low FPR obtained from experiments on a challenging and representative dataset reveals a great potential of our system in clinic.

References

1. World Health Organization: Latest global cancer data: Cancer burden rises to 19.3 million new cases and 10.0 million cancer deaths in 2020. International Agency for Research on Cancer. Geneva: World Health Organization (2020)

2. Chi, J., Walia, E., Babyn, P., Wang, J., Groot, G., Eramian, M.: Thyroid nodule classification in ultrasound images by fine-tuning deep convolutional neural network. J. Digit. Imaging **30**(4), 477–486 (2017)

3. Ma, J., Wu, F., Zhu, J., Xu, D., Kong, D.: A pre-trained convolutional neural network based method for thyroid nodule diagnosis. Ultrasonics **73**, 221–230 (2017)

4. Abdolali, F., Kapur, J., Jaremko, J.L., Noga, M., Hareendranathan, A.R., Punithakumar, K.: Automated thyroid nodule detection from ultrasound imaging using deep convolutional neural networks. Comput. Biol. Med. **122**, 103871 (2020)

5. Li, H., et al.: An improved deep learning approach for detection of thyroid papillary cancer in ultrasound images. Sci. Rep. **8**(1), 1–12 (2018)

6. Xie, S., Yu, J., Liu, T., Chang, Q., Niu, L., Sun, W.: Thyroid nodule detection in ultrasound images with convolutional neural networks. In: 14th IEEE Conference on Industrial Electronics and Applications (ICIEA), pp. 1442–1446 (2019)

7. Wang, L., et al.: Automatic thyroid nodule recognition and diagnosis in ultrasound imaging with the YOLOv2 neural network. World J. Surg. Oncol. **17**(1), 1–9 (2019)

8. Zhou, X., Koltun, V., Krähenbühl, P.: Tracking objects as points. In: Vedaldi, A., Bischof, H., Brox, T., Frahm, J.-M. (eds.) ECCV 2020. LNCS, vol. 12349, pp. 474–490. Springer, Cham (2020). https://doi.org/10.1007/978-3-030-58548-8_28

9. Hu, J., Shen, L., Sun, G.: Squeeze-and-excitation networks. In: 2018 IEEE/CVF Conference on Computer Vision and Pattern Recognition, pp. 7132–7141 (2018)

10. Sun, P., Jiang, Y., Xie, E., Yuan, Z., Wang, C., Luo, P.: OneNet: Towards End-to-End One-stage object detection. arXiv preprint arXiv:2012.05780 (2020)

11. Lin, T.Y., Goyal, P., Girshick, R., He, K., Dollár, P.: Focal loss for dense object detection. In: Proceedings of the IEEE International Conference on Computer Vision, pp. 2980–2988 (2017)

12. Rezatofighi, H., Tsoi, N., Gwak, J., Sadeghian, A., Reid, I., Savarese, S.: Generalized intersection over union: A metric and a loss for bounding box regression. In: Proceedings of the IEEE Conference on Computer Vision and Pattern Recognition, pp. 658–666 (2019)

13. Schroff, F., Kalenichenko, D., Philbin, J.: FaceNet: a unified embedding for face recognition and clustering. In: 2015 IEEE Conference on Computer Vision and Pattern Recognition (CVPR), pp. 815–823 (2015)

14. Tim, S., Ian, G., Wojciech, Z., Vicki, C.: Improved techniques for training GANs. In: Proceedings of the 30th International Conference on Neural Information Processing Systems (NIPS), pp. 2234–2242 (2016)

15. Tim, S., Milan, A., et al.: Mot16: a benchmark for multi-object tracking. arXiv preprint arXiv: 1603.00831 (2016)

16. Xingyi, Z., Dequan, W., Philipp, K.: Objects as Points. *CVPR* (2019)

17. Labit-Bonis, C., Thomas, J., Lerasle, F., Madrigal, F.: Fast tracking-by-detection of bus passengers with Siamese CNNs. In: 16th IEEE International Conference on Advanced Video and Signal Based Surveillance (AVSS), pp. 1–8 (2019)

TransBridge: A Lightweight Transformer for Left Ventricle Segmentation in Echocardiography

Kaizhong Deng[1], Yanda Meng[2], Dongxu Gao[2], Joshua Bridge[2],
Yaochun Shen[1], Gregory Lip[3], Yitian Zhao[4], and Yalin Zheng[2(✉)]

[1] Department of Electrical Engineering and Electronics, University of Liverpool,
Liverpool, UK
[2] Department of Eye and Vision Science, University of Liverpool, Liverpool, UK
yalin.zheng@liverpool.ac.uk
[3] Department of Cardiovascular and Metabolic Medicine, University of Liverpool,
Liverpool, UK
[4] Cixi Institute of Biomedical Engineering, Ningbo Institute of Materials Technology
and Engineering, Chinese Academy of Sciences, Ningbo, China

Abstract. Echocardiography is an essential diagnostic method to assess
cardiac functions. However, manually labelling the left ventricle region
on echocardiography images is time-consuming and subject to observer
bias. Therefore, it is vital to develop a high-performance and efficient
automatic assessment tool. Inspired by the success of the transformer
structure in vision tasks, we develop a lightweight model named 'Trans-
Bridge' for segmentation tasks. This hybrid framework combines a convolutional neural network (CNN) encoder-decoder structure and a transformer structure. The transformer layers bridge the CNN encoder and
decoder to fuse the multi-level features extracted by the CNN encoder, to
build global and inter-level dependencies. A new patch embedding layer
has been implemented using the dense patch division method and shuffled group convolution to reduce the excessive parameter number in the
embedding layer and the size of the token sequence. The model is evaluated on the EchoNet-Dynamic dataset for the left ventricle segmentation
task. The experimental results show that the total number of parameters is reduced by 78.7% compared to CoTr [22] and the Dice coefficient
reaches 91.4%, proving the structure's effectiveness.

Keywords: Echocardiography · Left ventricle segmentation ·
Lightweight transformer model · Parameter efficiency

1 Introduction

Cardiovascular disease has one of the highest mortality and morbidity rates
worldwide. Echocardiography imaging is essential for evaluating cardiac functions in clinical practice, such as left ventricular ejection fraction [16]. The left

© Springer Nature Switzerland AG 2021
J. A. Noble et al. (Eds.): ASMUS 2021, LNCS 12967, pp. 63–72, 2021.
https://doi.org/10.1007/978-3-030-87583-1_7

ventricular ejection fraction assessment is usually performed by comparing the left ventricular volume at end-systolic and end-diastolic frames. Manual annotation of the left ventricular region is a time-consuming and human-dependent step, resulting in high inter-observer variance and limited precision [8,11]. Hence, it is vital to develop an automatic segmentation algorithm of the left ventricle in echocardiographic images. Some machine learning methods have been proposed, such as Structured Random Forest [9] and dynamic appearance model [7]. However, they are either based on hand-crafted features or not sufficiently robust. Recent research interest moves to the deep learning methods that will avoid hand-crafted features and are robust enough. Several models using distinct network structures have shown promising performance [10,12,17], while [12] provides a comprehensive review of the recent methods. One of the limitations of these methods is the large model size that is not efficient to use.

Related Works. The development of deep learning methods and approaches [2,3,14,15,19] has led to improvements in biomedical image segmentation tasks. For example, U-Net [19] uses encoder-decoder architecture with the skip-connection to extract features from multiple scales and recover them to the original scale. It has been shown that the U-Net reaches good accuracy on left ventricle segmentation [10]. The residual connection in ResNet [5] improves the accuracy of the CNN by constructing a clean identity mapping path to ease optimization [6], and ResUNet [21] employs this technique in the U-Net structure. DeepLabV3 [2] uses dilated convolutions to increase the receptive field so that the model can catch dependency at a longer distance. It has been shown that DeepLabV3 can reach a remarkable performance on the left ventricle segmentation task [17]. In a recent study, the transformer model is introduced to break through the limitation of locality from convolution operators to build the global dependency. The Vision Transformer [4] is a pure Transformer model in image recognition tasks with state-of-the-art performance. The transformer model combined with CNN structure has also shown great potential in the image segmentation task [1,22,24]. However, the drawback of introducing transformer structures is the significant increase in the number of parameters. Therefore, it is necessary to design a lightweight transformer model to utilize its high performance on vision tasks. For example, works on reducing parameter number in CNNs and transformers by applying shuffle algorithm have been proposed in [13,23]. The Sandwich parameter sharing the transformer encoder structure has also been discussed [18]. Therefore, building an efficient and training-friendly model should also be a crucial criterion of the deep learning model.

Our Contributions. In our works, the patch embedding before the Transformer structures are re-designed using the shuffling layer and group convolutions to reduce the excessive parameter number and token numbers. Sandwich parameter sharing was used to minimize the transformer parameters [18]. We propose the TransBridge, a lightweight hybrid model using the transformer and the CNN structure for left ventricle segmentation in echocardiography.

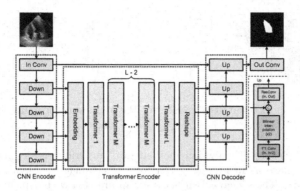

Fig. 1. TransBridge: Downsampling block in CNN Encoder and upsampling block in CNN decoder. The transformer bridges the CNN encoder and decoder to model inter-feature level dependency. The Sandwich parameter sharing mechanism allows parameters shared in all the middle layers except for the beginning and the end.

2   Methods

CNN Encoder. The CNN encoder is used to extract features efficiently to obtain high abstract level features, saving time for the transformer encoder to focus its attention on low-level features. The CNN encoder adopts the U-Net encoder structure that cascades convolution layers and downsamples the resulting features between each block [19], shown in Fig. 1. The downsampling layer comprises a max-pooling operation to downsample the feature map size and a residual double convolution block. The residual block contains two BN-ReLU-Conv layers and a 1×1 Conv for identity mapping. In addition, the Pre-activation residual block can result in easier training [6]. Assuming the input image is of size of (H, W), the extracted feature maps can be expressed as $\{x\}_l \in \mathbb{R}^{(C \times l) \times \frac{H}{T} \times \frac{W}{T}}, 1 \leq l \leq L$. Considering the efficiency of the model, the feature maps from the first CNN encoder layer are not used for the transformer encoder layer but directly skip-connect to the CNN decoder layer at the same level to retain the low-level features and reduce the cost of attention.

Patch Division. In the transformer-based vision task, such as ViT [4] and SeTr [24], the input of the transformer encoder layers is embedded patch sequence. In the embedding layer, shown in Fig. 2(a), the input image $x \in \mathbb{R}^{C \times H \times W}$ is equally divided into patches. Every patch is flattened to a 1-dimensional vector so that the patch sequence becomes $p \in \mathbb{R}^{N \times D_0}$, where the number of patches is represented as $N = \frac{H}{P} \times \frac{W}{P}$ and the vector size is represented as $D_0 = C \times P \times P$.

In order to embed the multi-channel feature maps, channels are split into several groups and treated independently. The patch is divided with a fixed size of (P, P), and D channels in each group are attributed to the patch, as shown in Fig. 2(b). Therefore, the sequence of a patch of feature maps is in the form of a sequence of dense patches $p \in \mathbb{R}^{\frac{C}{D} \times \frac{H}{P} \times \frac{W}{P} \times D \times P \times P}$. After flattening

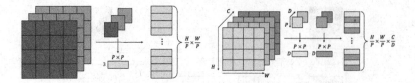

(a) Patch division for RGB image   (b) Patch division for multi-channel
feature maps

Fig. 2. Patch division: The division method in TransBridge is designed for multi-channel feature maps as in Fig. 2(b). The total C channels are divided into several D channel groups. Then, channels in each group are treated independently. Finally, those D-channeled patches are flattened into vectors and concatenated.

the dense patches to vector, the feature map $x \in \mathbb{R}^{C \times H \times W}$ is transformed into a dense flattened patch sequence of $z_d \in \mathbb{R}^{M \times D_d}$, where the vector length is $D_d = D \times P \times P$, and the total number is $M = \frac{C}{D} \times \frac{H}{P} \times \frac{W}{P}$. As feature maps from the different levels have different channel sizes and spatial dimension, the number of token M is different in each level. However, the input size of the patch D_d is the same among all sequence so that the token sequence is $\{z_d\}_l \in \mathrm{R}^{M_l \times D_d}$ and the l denotes the level of features.

Length Shortening. The length of the token sequence is shortened before patch embedding. The token length is crucial because the complexity of the transformer encoder layer is sensitive to the sequence length. In this design, a shuffling layer and 1×1 group convolution are applied to shorten the length. First, in the shuffling layer, as shown in Fig. 3, all the four token sequences are divided into G groups individually through the channel dimension and all divided sequences from different feature levels are shuffled according to the group number to rearrange the group division so that each new group contains an element from each level. Next, sequences are concatenated through the channel dimension and conduct a 1×1 convolution in a group of G to compress the channel number to N to shorten the total sequence length.

Transformer Encoder. Before feeding into the transformer encoder, patch and positional embedding are required to pre-process the patch sequence. A trainable linear layer projects the token vector from its length D_d to the hidden size D_h of the transformer encoder to obtain the patch embedding as shown in Eq.(1). Next, a trainable positional embedding layer is added to the patch embedding to retain the spatial information that the transformer encoder layer cannot model.

$$z_0 = \left[z_h^1 E; z_h^2 E; \ldots; z_h^N E \right] + E_{pos}, E \in R^{D_d \times D_h}, E_{pos} \in R^{N \times D_h} \tag{1}$$

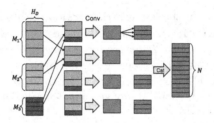

Fig. 3. Shuffling layer and group convolution: tokens from different feature channels are firstly split into groups and shuffled; A group convolution is applied to the grouped tokens to shorten the token sequence length from $M_1 + M_2 + M_3$ to N. All the tokens are concatenated together at the final stage. For the demonstration purpose, the level of features L is set to 3 and the group number G is 4.

Every single layer of the transformer encoder consists of Multihead Self-Attention (MSA) blocks and Multi-Layer Perceptron (MLP) blocks, shown in Eq. (2) and (3). A residual connection bypasses each block to form an identity mapping and a layer normalization operator is inserted in the front of each block. In addition, to increase the parameter efficiency, the parameter is shared in the Sandwich mode [18], which shares the parameters of all $L - 2$ middle layers, except the beginning and the ending layer in this L layer transformer encoder.

$$z'_l = MSA\left(LN\left(z_{l-1}\right)\right) + z_{l-1} \tag{2}$$

$$z_l = MLP\left(LN\left(z'_{l-1}\right)\right) + z'_{l-1} \tag{3}$$

The token sequence will be expanded and rearranged by reversing the length compression and patch division back to feature maps with the original dimension. During the rearranging, the shuffling process is not applied because the channel dimension has already been mixed.

CNN Decoder. The CNN decoder absorbs feature maps from the transformer encoder and recovers them to the original size. For example, in the upsampling block, the feature maps from the previous decoder layer use 1×1 convolutions to match the channel numbers to half of the desired input channel number. Then its height and width are doubled by bilinear interpolation. Next, the resulted planes are concatenated with the feature maps from the transformer encoders to feed into a residual block to refine the feature maps. Finally, the output block will fuse the resulted planes into a one-dimensional segmentation map to output it as the final prediction.

3 Experiments

Dataset. EchoNet-Dynamic dataset is a large public dataset with apical four-chamber two-dimensional echocardiographs [17]. For each video, an end-systole and an end-diastole frame were selected for the analysis. Expert sonographers and cardiologists annotate the left ventricle region during the standard clinical workflow. Among the 20,048 images, 14,920 images were used for training, 2,576 images for validating, and 2,552 images for testing. The end-systolic and end-diastolic frame of the same subject were placed in the same group. All the images were resized to 112 × 112 pixels and converted to grayscale. The training set was shuffled in each epoch to avoid any specific class distribution in each batch.

Implementation Details. The proposed model was implemented with two scales: Base ('TransBridge-B') and Large ('TransBridge-L'). To better compare with the TransBridge, the CoTr [22] was implemented with the original Vaswani Transformer instead of the Deformable Transformer and built in the base scale. The differences between the two scales of the TransBridge are CNN channel number, transformer hidden size, and transformer MLP intermediate layer size, shown in Table 1. The number of CNN feature levels fed to the Transformer, L, is set to 4. The patch size was set to $(7, 7)$, and the grouping factor G was set to 8 so that there were at least two groups in each feature level for the shuffling. The transformer encoder layer has six layers and is split into four heads in the self-attention layer. The parameter number of TransBridge-B has been reduced by 78.7% compared with the CoTr model. Meanwhile, the number of parameters of the embedding layer has been reduced from 12.07M in CoTr to 0.17M in TransBridge, which is 1.4% of the normal embedding layer. The UNet and the ResUNet have also been implemented as references. The ResUNet has the CNN structure but without the transformer encoder layer in TransBridge.

Table 1. The configurations of the evaluated models

Method	Total Param	Embedding Layer Param	CNN Structure	Transformer Structure	
			L1 Channel Number	Hidden size	MLP size
CoTr	16.39M	12.07M	16	256	256
TransBridge-B	3.49M	0.17M	16	256	256
TransBridge-L	11.3M	0.23M	32	392	512
UNet	7.25M	-	32	-	-
ResUNet	7.6M	-	32	-	-

The model was trained on an Nvidia Tesla P100 GPU with a batch size of 8. The running GPU memory of our model can be limited to approximately 2 GB. All the models were trained with an RMSprop optimizer with learning rate of 1e−4, momentum of 0.9, and a weight decay of 1e−8 for 15 epochs. Each epoch contains 20 steps, and each step has 93 iterations. The learning rate dropped to 10% of its original value if there is no further improvement in 10 steps. Binary cross-entropy loss is used to train the model and the Dice loss is used for validation.

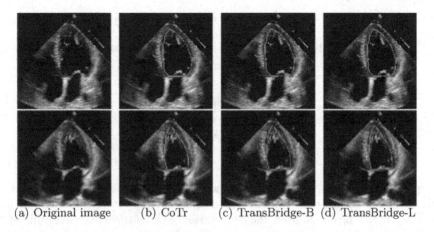

(a) Original image   (b) CoTr   (c) TransBridge-B (d) TransBridge-L

Fig. 4. Segmentation Results Visualization: The ground truth is labeled with a green line, while the segmentation boundary from each model is in red. (Color figure online)

4 Results

Comparison Between Models. The performance of the TransBridge in two scales is compared with other methods on the left ventricle segmentation task. The segmentation are divided into two groups based on the heart contraction stage, either end-systolic or end-diastolic. Dice coefficient and Hausdorff distance are used to evaluate the segmentation quality.

Table 2. Comparing the segmentation results of: TransBridge-B (ours), TransBridge-L (ours), CoTr [22] are trained on the dataset. In addition, the results of the UNet and DeepLabV3 are cited from [10] and [17] respectively.

Method	Hausdorff distance					Dice (in %)				
	End-systolic		End-diastolic		Average	End-systolic		End-diastolic		Average
	Mean	Std	Mean	Std		Mean	Std	Mean	Std	
UNet [10]	-	-	-	-	7.3	-	-	-	-	89.6
DeepLabV3 [17]	-	-	-	-	-	90.3	-	92.7	-	91.5
UNet	6.506	5.977	6.017	4.405	6.262	82.50	0.078	87.63	0.054	85.07
ResUNet	4.175	5.403	3.725	5.403	3.950	91.17	0.048	93.51	0.034	92.34
CoTr	4.699	5.838	4.201	3.652	4.450	89.87	0.061	92.71	0.042	91.29
TransBridge-B	4.633	5.853	4.184	3.757	4.409	90.01	0.057	92.76	0.037	91.39
TransBridge-L	4.411	5.528	3.959	3.346	4.185	90.24	0.058	93.04	0.035	91.64

The testing results are shown in Table 2 and Fig. 4. Comparing the TransBridge-B and TransBridge-L with CoTr, improvements are made on the Dice coefficient (91.69% and 91.39% vs. 91.29%) and Hausdorff distance (4.185 and 4.409 vs. 4.450). In particular, TransBridge-B has only 21.3% parameters of CoTr, so it is more lightweight and efficient. Meanwhile, UNet, ResUNet, and

Table 3. Ablation test with each structure configuration

The first layer skip connection	CNN-block	Sandwich sharing	Dice (in %)
No	Conv	No	90.7
No	ResConv	Yes	90.7
Yes	Conv	No	89.9
Yes	ResConv	No	90.2
Yes	ResConv	Yes	91.0

DeepLabV3 have also been compared with the TransBridge models. In previous work [10], the UNet is evaluated on a small dataset with 1000 images. When training on this larger dataset, the large variance on features makes it difficult to perform as well as in the smaller dataset, and the further increment on its width cannot contribute to better accuracy. However, after introducing the Residual block, the accuracy of ResUNet has improved compared to UNet, exceeding the performance of DeepLabV3 and TransBridge. The reason for this might be that for this specific dataset, the image size is relatively small and the LV geometry is simple to segment, so there is no need for complicated models.

Ablation Test. In the ablation test, the model is trained until early converged, and it takes no more than five epochs for the validation loss to converge with tolerance less than 0.001. The results show that almost all the design changes can improve the overall performance, shown in Table 3. Sandwich sharing can make the most significant progress. Using the residual block instead of simple convolutions cannot make sufficient progress but it can avoid gradient vanishing. The skip connection of the first layer introduce low-level features, and its effect on the overall performance might depend on the presence of the other two features.

5   Discussion

The proposed TransBridge shows excellent potential for the left ventricle segmentation task. This lightweight design reduces the parameter by 78.7% while achieving a Dice score of 91.4%. In addition, the group and shuffling embedding can facilitate the information exchange in different feature levels and channels with fewer parameters. However, compared to the pure CNN structure, the transformer is not easy to train and attain competitive performance. It is sensitive to the dataset and hyperparameters, demanding extensive large-scale empirical trials to achieve the best performance [20]. Therefore, more sophisticated hyper-parameter tuning could further enhance the performance of the model.

6   Conclusion

This paper has proposed TransBridge, an efficient lightweight model that combines the CNN and transformer architecture for the LV segmentation task. The proposed shuffling layer and group convolution for patch embedding significantly reduces the total number of parameters by 78.7% and efficiently utilizes the transformer's power to cooperate with CNN. The model has been evaluated on the largest public echocardiography dataset, and the results confirm its effectiveness. In the future, the proposed model can be used as a powerful tool to support the management of cardiovascular diseases.

References

1. Chen, J., et al.: TransUNet: Transformers make strong encoders for medical image segmentation, February 2021
2. Chen, L.C., Papandreou, G., Schroff, F., Adam, H.: Rethinking atrous convolution for semantic image segmentation (2017)
3. Chen, X., Williams, B.M., Vallabhaneni, S.R., Czanner, G., Williams, R., Zheng, Y.: Learning active contour models for medical image segmentation. In: Proceedings of the IEEE/CVF Conference on Computer Vision and Pattern Recognition, pp. 11632–11640 (2019)
4. Dosovitskiy, A., et al.: An image is worth 16×16 words: transformers for image recognition at scale, October 2020
5. He, K., Zhang, X., Ren, S., Sun, J.: Deep residual learning for image recognition. In: Proceedings of the IEEE Conference on Computer Vision and Pattern Recognition (CVPR), June 2016
6. He, K., Zhang, X., Ren, S., Sun, J.: Identity mappings in deep residual networks. In: Leibe, B., Matas, J., Sebe, N., Welling, M. (eds.) ECCV 2016. LNCS, vol. 9908, pp. 630–645. Springer, Cham (2016). https://doi.org/10.1007/978-3-319-46493-0_38
7. Huang, X., et al.: Contour tracking in echocardiographic sequences via sparse representation and dictionary learning. Med. Image Anal. 18(2), 253–271 (2014)
8. Lang, R.M., et al.: Recommendations for cardiac chamber quantification by echocardiography in adults: an update from the american society of echocardiography and the european association of cardiovascular imaging. Eur. Heart J. Cardiovascular Imag. 16(3), 233–271 (2015). https://doi.org/10.1093/ehjci/jev014
9. Leclerc, S., Grenier, T., Espinosa, F., Bernard, O.: A fully automatic and multi-structural segmentation of the left ventricle and the myocardium on highly heterogeneous 2D echocardiographic data. In: 2017 IEEE International Ultrasonics Symposium (IUS), pp. 1–4 (2017). https://doi.org/10.1109/ULTSYM.2017.8092797
10. Leclerc, S., et al.: Deep learning applied to multi-structure segmentation in 2D echocardiography: a preliminary investigation of the required database size. In: 2018 IEEE International Ultrasonics Symposium (IUS), pp. 1–4 (2018). https://doi.org/10.1109/ULTSYM.2018.8580136
11. Leclerc, S., et al.: Deep learning for segmentation using an open large-scale dataset in 2D echocardiography. IEEE Trans. Med. Imag. 38(9), 2198–2210 (2019). https://doi.org/10.1109/TMI.2019.2900516
12. Li, M., et al.: Unified model for interpreting multi-view echocardiographic sequences without temporal information. Appl. Soft Comput. 88, 106049 (2020)
13. Mehta, S., Ghazvininejad, M., Iyer, S., Zettlemoyer, L., Hajishirzi, H.: Delight: Deep and light-weight transformer, August 2020
14. Meng, Y., et al.: Regression of instance boundary by aggregated CNN and GCN. In: Vedaldi, A., Bischof, H., Brox, T., Frahm, J.-M. (eds.) ECCV 2020. LNCS, vol. 12353, pp. 190–207. Springer, Cham (2020). https://doi.org/10.1007/978-3-030-58598-3_12

15. Meng, Y., et al.: CNN-GCN aggregation enabled boundary regression for biomedical image segmentation. In: Martel, A.L., et al. (eds.) MICCAI 2020. LNCS, vol. 12264, pp. 352–362. Springer, Cham (2020). https://doi.org/10.1007/978-3-030-59719-1_35

16. Oktay, O., et al.: Anatomically constrained neural networks (ACNNs): application to cardiac image enhancement and segmentation. IEEE Trans. Med. Imag. **37**(2), 384–395 (2018). https://doi.org/10.1109/TMI.2017.2743464

17. Ouyang, D., et al.: Video-based AI for beat-to-beat assessment of cardiac function. Nature **580**(7802), 252–256 (2020). https://doi.org/10.1038/s41586-020-2145-8

18. Reid, M., Marrese-Taylor, E., Matsuo, Y.: Subformer: Exploring weight sharing for parameter efficiency in generative transformers (2021)

19. Ronneberger, O., Fischer, P., Brox, T.: U-Net: convolutional networks for biomedical image segmentation. In: Navab, N., Hornegger, J., Wells, W.M., Frangi, A.F. (eds.) MICCAI 2015. LNCS, vol. 9351, pp. 234–241. Springer, Cham (2015). https://doi.org/10.1007/978-3-319-24574-4_28

20. Steiner, A., Kolesnikov, A., Zhai, X., Wightman, R., Uszkoreit, J., Beyer, L.: How to train your vit? data, augmentation, and regularization in vision transformers (2021)

21. Xiao, X., Lian, S., Luo, Z., Li, S.: Weighted Res-UNet for high-quality retina vessel segmentation (2018). https://doi.org/10.1109/ITME.2018.00080

22. Xie, Y., Zhang, J., Shen, C., Xia, Y.: CoTr: Efficiently bridging CNN and transformer for 3D medical image segmentation, March 2021

23. Yang, Q.L.Z.Y.B.: SA-Net: Shuffle attention for deep convolutional neural networks, January 2021

24. Zheng, S., et al.: Rethinking semantic segmentation from a sequence-to-sequence perspective with transformers. In: Proceedings of the IEEE/CVF Conference on Computer Vision and Pattern Recognition (CVPR), pp. 6881–6890, June 2021

Registration, Guidance and Robotics

Adversarial Affine Registration for Real-Time Intraoperative Registration of 3-D US-US for Brain Shift Correction

Marek Wodzinski[✉][iD] and Andrzej Skalski[iD]

Department of Measurement and Electronics, AGH University of Science and Technology, Krakow, Poland
wodzinski@agh.edu.pl

Abstract. One of the most frequent tumors in the central nervous system is glioma. The high-grade gliomas grow relatively fast and eventually lead to death. The tumor resection improves the survival rate. However, an accurate image-guidance is necessary during the surgery. The problem may be addressed by image registration. There are three main challenges: (i) the registration must be performed in real-time, (ii) the tumor resection results in missing data that strongly influence the similarity measure, and (iii) the quality of ultrasonography images. In this work, we propose a solution based on generative adversarial networks. The generator network calculates the affine transformation while the discriminator network learns the similarity measure. The ground-truth for the discriminator is defined by calculating the best possible affine transformation between the anatomical landmarks. This approach allows real-time registration during the inference and does not require defining the similarity measure that takes into account the missing data. The work is evaluated using the RESECT database. The dataset consists of 17 US-US pairs acquired before, during, and after the surgery. The target registration error is the main evaluation criteria. We show that the proposed method achieves results comparable to the state-of-the-art while registering the images in real-time. The proposed method may be useful for the real-time intraoperative registration addressing the brain shift correction.

Keywords: Image registration · Deep learning · GANs · RESECT · Ultrasonography · Glioma

1 Introduction

Glioma is one of the most frequently occurring tumors in the central nervous system [1]. The high-grade gliomas grow relatively fast and eventually lead to death. The tumor resection improves the survival rate [2–4]. An accurate image-guidance is helpful during the surgery. The tissue deformations resulting from the tumor removal, insertion of surgical instruments, or pressure changes need to be addressed in real-time.

© Springer Nature Switzerland AG 2021
J. A. Noble et al. (Eds.): ASMUS 2021, LNCS 12967, pp. 75–84, 2021.
https://doi.org/10.1007/978-3-030-87583-1_8

A promising imaging technique for addressing tissue deformations is intra-operative ultrasonography (US). Its flexibility, portability, and low cost make it especially useful in clinical practice. Moreover, the tumors are easily distinguishable in the US images from other neighboring tissues. Therefore, it would be beneficial to propose a real-time image registration method that aligns the pre-operative US volumes to the intra- or post-operative ones.

Unfortunately, the problem is very challenging. There are two main difficulties related to 3-D US-US registration: (i) the requirement of the real-time registration, and (ii) the missing data due to the tumor resection, surgical instruments insertion, varying position of the probe, and (iii) the quality of US images. The registration should be ideally performed with a frequency equal to the frame rate. This is usually impossible using the classical, iterative methods. The state-of-the-art techniques with a GPU implementation achieve at most several Hz [5]. The missing data results in difficulties with defining the correct similarity measure. Even though similarity measures like correlation ratio or LC^2 are quite useful in aligning the US volumes [6,7], they fail when the region of interest is being resected. The US images are strongly affected by the speckle noise, and the contrast is significantly worse than in, e.g., magnetic resonance images or computed tomography volumes. Therefore, the ideal image registration method for intraoperative US volumes' alignment should be fast, resistant to tissue resection changes, and robust to the US's properties.

There are numerous contributions to the registration of 3-D US-US volumes [6]. Perhaps the most influential work is the affine registration based on the block matching algorithm implemented on GPU [5]. The authors achieved real-time processing. However, the method was evaluated using volumes with a relatively similar field of view and without missing structures. The researchers re-evaluated the method on the RESECT database replacing the original formulation based on the affine transformation into the rigid transformation [8,9]. Another contribution proposed a feature-based algorithm [10]. The method consists of automatic feature extraction and matching followed by a dense displacement field interpolation. In [11] authors proposed a feature-based contribution that involves the calculation of 3-D SIFT features followed by the dense displacement field interpolation by thin-plate splines. Even though the affine registration is perfectly justified, the following thin-plate splines interpolation is controversial. Since the SIFT features are not extracted close to the tumor (there are no mutual correspondences), the interpolation is undefined in these regions. It may result in an incorrect registration from the medical point of view. Thus, an additional evaluation is necessary for the nonrigid method. Another interesting contribution is segmentation-based method [12]. The authors apply Euclidean Distance Transform (EDT) to the segmented structures. Then, they match the distance images using the normalized gradient fields as the similarity measure, together with the curvature regularization. The method achieves the best results on the RESECT database. However, it requires the segmentation mask that needs to be defined before and during the surgery.

The challenges related to the registration of US volumes may be addressed by the novel deep learning (DL) algorithms, more specifically, the generative adversarial networks (GANs) [13,14]. The DL-based image registration methods are usually much faster than the classical, iterative algorithms [15,16]. The computational complexity is transferred to the training phase that is performed on servers dedicated to parallel computations. Then, during the inference, the registration may be performed in real-time. On the other hand, GANs enable the possibility to make the similarity measure learnable. The discriminator is being trained to distinguish between correctly aligned and misaligned images. As a result, the proper training may lead to learning only the features with correspondence in both the images and ignoring the missing regions.

Contribution: The work presents an image registration method dedicated to the intraoperative, affine US-US registration. We propose a GAN-based solution. The generator is responsible for calculating the affine transformation while the discriminator implicitly learns the similarity measure used to guide the generator during training. The presented method does not need the predefined similarity measure neither the segmentation masks. Moreover, the time required for the network inference is low, thus enabling real-time registration. The proposed method solves the most significant challenges related to the registration of intraoperative US volumes.

2 Methods

2.1 Affine Registration

The proposed method consists of a generator responsible for calculating the affine transformation and a discriminator that guides the generator during training. The generator outputs directly the affine transformation (as a 2×3 matrix). The architectures, together with the method overview, are shown in Fig. 1. The input images were downsampled to a lower resolution ($2\times$) to decrease the GPU memory consumption.

The training is performed by repeating the following steps:

1. Forward through the discriminator the correctly aligned source/target pair, calculate the objective function $(O(M(\boldsymbol{x}), F(\boldsymbol{x})) = -log(D(M(\boldsymbol{x}), F(\boldsymbol{x}))))$ and update the discriminator weights.
2. Forward through the generator the misaligned source/target pair, transform the source, pass the transformed source/target pair to the discriminator, calculate the objective function $(O(M(\boldsymbol{x}), F(\boldsymbol{x}), u(\boldsymbol{x})) = -log(1 - D((M \circ u)(\boldsymbol{x}), F(\boldsymbol{x}))))$ and update the discriminator weights.
3. Forward through the generator the misaligned source/target pair, transform the source, pass the transformed source/target pair to the discriminator, calculate the objective function $(O(M(\boldsymbol{x}), F(\boldsymbol{x}), u(\boldsymbol{x})) = -log(D((M \circ u)(\boldsymbol{x}), F(\boldsymbol{x}))))$ and update the generator weights.

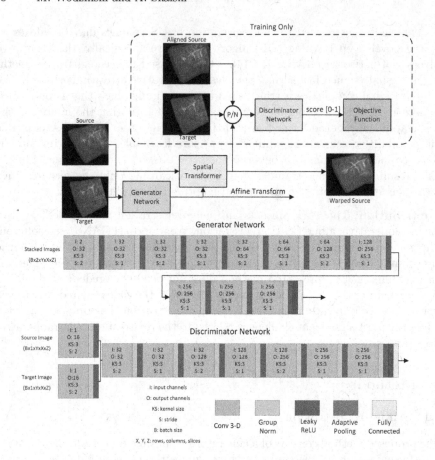

Fig. 1. Visualization of the training scheme and the generator/discriminator networks.

The $M(\boldsymbol{x}), F(\boldsymbol{x})$ denote the moving and fixed image respectively, $u(\boldsymbol{x})$ is the sampling grid defined by the calculated affine transform, $D(\cdot, \cdot)$ denotes the output from the discriminator, $\circ$ is the composition.

The positive cases are defined as image pairs aligned with the best possible affine transformation (offline, prior to the network training). The transformations are calculated by minimizing the mean Euclidean distance between the anatomical landmarks.

The training requires several tricks to improve the convergence. Without them, the discriminator automatically overfits the training data, and the generator is unable to learn how to calculate correct transformations. They are as follows:

1. Strongly augment the training data by random affine transformations. Both the positive and negative pairs should be randomly augmented. Without the augmentation, the discriminator instantly overfits.

2. Apply the Gaussian filtering before the registration and warp the images using the same trilinear interpolation. This prevents the discriminator from learning the patterns in the speckle noise or the interpolation artifacts.
3. Pre-train the generator and discriminator using another, larger dataset. The networks were pretrained using resampled and randomly transformed dataset for the registration of abdominal organs [17,18].
4. Schedule the discriminator and generator epochs separately. In contrast to the GANs used for the image synthesis, the GANs for the image registration require a larger number of epochs for the generator than for the discriminator, especially with a small training set.
5. The positive and negative loss should get similar values within a few initial epochs. Otherwise, the discriminator is overfitted, and a fresh start is required.

2.2 Dataset and Experimental Setup

The dataset consists of 17 US-US cases from the RESECT database [8]. The images were acquired: (i) before the surgery, (ii) during the surgery, and (iii) after the surgery. Two experienced medical experts annotated the ground-truth anatomical landmarks used for the evaluation. The mean difference between the two annotators is equal to 0.27 mm. The pre- to post- resection pairs contain 10–17 landmarks, while the pre- to during-resection pairs have 16–34 landmarks. The initial TRE for each case is presented in Table 1 and the distribution is shown in Fig. 2. The dataset is openly available. Its full description is available in the RESECT article [8].

The dataset was divided into six-folds. In each fold, 2–3 pairs were considered the test cases and were not used during training. The remaining pairs were considered as training pairs. One pair within the training set was used as a validation pair to monitor discriminator convergence. We accumulated the results from the six folds by calculating the transformation for the test cases. All the experiments shared the same learning rate (0.0002). The number of epochs was different for each fold since we performed training manually to ensure convergence. The before-during pairs were trained separately from the before-after pairs.

3 Results

The target registration error (TRE) is the main evaluation criteria. It is defined as the Euclidean distance between the landmarks annotated in the moving image and the transformed landmarks from the fixed image. The cumulative histograms presenting the TRE before, after, and for the ideal affine registration are shown in Fig. 2. Table 1 presents the quantitative results for each case. It also contains the results reported by other methods. An exemplary visual assessment of the registration is shown in Fig. 3. The average registration time, defined as the time required to calculate the affine transformation during the inference, is equal to 42 ms (using RTX 2080 Ti).

Table 1. Table presenting the TRE compared to other methods evaluated on the RESECT dataset. The method based on block-matching algorithm [9] did not report the ranges. The table presents the registration results for the registration of images acquired before and after the surgery. The results for images acquired before and during the surgery are omitted because other researchers do not report them. Please note that the block-matching method [9] reports the results in different format than other contributions. *Ground-Truth* defines the best possible affine transformation.

Case	Initial	*Ground-Truth*	**Proposed (GAN)**	Block-Matching [9]
1	5.80 (3.62–7.22)	0.97 (0.24–2.52)	1.22 (0.36–2.55)	1.34
2	3.65 (1.71–6.72)	1.30 (0.21–2.33)	**1.71 (0.69–3.08)**	4.63
3	2.91 (1.53–4.30)	0.68 (0.23–1.39)	**0.95 (0.52–1.65)**	1.34
4	2.22 (1.25–2.94)	0.56 (0.25–1.07)	1.18 (0.66–1.93)	0.91
6	2.12 (0.75–3.82)	1.31 (0.56–2.59)	**1.48 (0.58–2.58)**	3.07
7	3.62 (1.19–5.93)	1.45 (0.41–3.45)	2.31 (0.61–3.31)	2.61
12	3.97 (2.58–6.35)	1.06 (0.51–2.93)	1.69 (0.85–2.60)	1.65
14	0.63 (0.17–1.76)	0.44 (0.17–0.76)	1.34 (0.21–2.38)	0.60
15	1.63 (0.62–2.69)	0.71 (0.26–1.73)	1.37 (0.46–2.61)	0.90
16	3.13 (0.82–5.41)	1.07 (0.30–2.19)	1.43 (0.78–2.63)	3.12
17	5.71 (4.25–8.03)	0.95 (0.30–2.17)	2.18 (0.84–3.84)	1.83
18	5.29 (2.94–9.26)	1.18 (0.57–2.19)	1.60 (0.53–2.58)	2.29
19	2.05 (0.43–3.24)	0.88 (0.23–2.48)	1.82 (0.42–2.86)	1.81
21	3.35 (2.34–5.64)	0.82 (0.35–1.52)	1.10 (0.44–1.55)	1.87
24	2.61 (1.96–3.41)	0.65 (0.14–1.56)	0.91 (0.32–2.03)	1.06
25	7.61 (6.40–10.25)	0.88 (0.38–1.55)	2.56 (1.91–3.38)	2.84
27	3.98 (3.09–4.82)	0.46 (0.13–0.61)	1.07 (0.41–1.85)	0.71
Mean	**3.55 (2.10–5.40)**	**0.91 (0.31–1.94)**	**1.51 (0.62–2.55)**	**1.92 ± 1.04**

Case	EDT [12]	SIFT (Affine) [10]	SIFT (Nonrigid) [10]	
1	**1.05 (0.28–2.48)**	1.64 (0.14–3.71)	1.48 (0.17–3.51)	
2	2.32 (0.42–4.16)	2.63 (0.85–5.14)	2.62 (0.62–4.89)	
3	1.39 (0.55–2.24)	1.19 (0.64–2.50)	1.04 (0.62–1.52)	
4	**0.81 (0.25–1.80)**	0.92 (0.22–1.50)	0.83 (0.27–1.53)	
6	1.62 (0.39–4.65)	1.97 (0.51–3.73)	1.55 (0.67–2.88)	
7	**1.25 (0.25–3.15)**	2.59 (0.84–5.11)	2.38 (0.45–4.33)	
12	**0.87 (0.20–1.82)**	1.21 (0.24–3.78)	1.20 (0.44–3.09)	
14	0.62 (0.32–1.10)	0.53 (0.08–1.21)	**0.53 (0.18–1.18)**	
15	0.80 (0.27–1.81)	0.79 (0.26–2.42)	**0.74 (0.29–2.31)**	
16	**1.26 (0.22–3.91)**	1.97 (0.48–4.25)	1.94 (0.20–3.84)	
17	**1.51 (0.47–5.59)**	1.97 (0.94–4.72)	1.99 (0.21–4.51)	
18	**1.53 (0.30–3.61)**	1.71 (0.71–3.36)	1.69 (0.58–3.03)	
19	**1.60 (0.39–3.45)**	2.46 (0.67–5.19)	2.78 (0.65–5.04)	
21	1.82 (0.25–5.12)	1.23 (0.49–3.57)	**1.07 (0.56–3.20)**	
24	**0.90 (0.24–2.33)**	1.32 (0.44–2.63)	1.35 (0.35–2.24)	
25	**1.00 (0.30–2.44)**	1.51 (0.35–3.87)	1.24 (0.21–3.57)	
27	1.24 (0.35–2.74)	**0.48 (0.05–0.96)**	0.83 (0.20–0.93)	
Mean	**1.27 ± 0.44**	**1.54 (0.47–3.39)**	**1.49 (0.39–3.04)**	

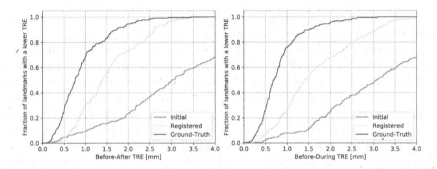

Fig. 2. The cumulative histogram presenting the TRE before and after the registration. The histogram presents also the results for the best possible affine transformation (denoted as "Ground-Truth").

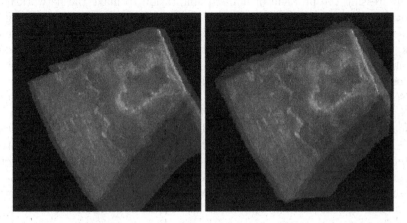

Fig. 3. An exemplary visualization of the images before (left) and after (right) the affine registration (Case 27). The target image is shown in green, the source image is presented in magenta. (Color figure online)

4   Discussion and Conclusion

The proposed method is robust and decreases the TRE for most cases within the RESECT database. The results are comparable to the block-matching algorithm, yet slightly worse compared to the nonrigid methods. This could be expected since the method calculates just the affine transformation. Interestingly, the method achieves better results in several cases than the nonrigid methods. On the other hand, for one case, the TRE even increases. The generator's inference time is relatively low and enables real-time US-US registration, up to 24 Hz. The average processing time is not directly discussed in the related works, yet state-of-the-art GPU implementations hardly achieve 8 Hz [5].

The method has a few limitations. Perhaps the most important one is connected with the stability and difficulty of training. The training of GANs is inherently unstable and, with such a small dataset, requires much manual interaction. The networks cannot be trained fully automatically since the discriminator would easily overfit the training data. As a result, one needs to control the training convergence and epoch scheduling. However, this approach is still more promising compared to supervised approach that with so small dataset instantly overfits. The training set must be prepared with care. The positive cases should contain the same interpolation artifacts and speckle noise characteristics as the negative cases. Otherwise, the discriminator learns features not related to the alignment quality.

The further research will involve several topics: (i) improving the training stability, (ii) enabling the use of GANs for the nonrigid US-US registration, (iii) addressing the problem of domain adaptation to make the network useful for other datasets. One could improve the training stability by enlarging the dataset. However, this is a huge task for physicians beyond the scope of this work. Another solution may be connected with decoupling the training of the discriminator and the generator. Perhaps training the discriminator to minimize the loss with respect to the TRE directly would be a good approach. However, it also requires a substantially larger dataset. The nonrigid registration using GANs is more challenging since it requires a perfectly aligned ground-truth that is usually impossible to acquire. One could argue that the non-ideal ground-truth could be created by, e.g., thin-plate splines interpolation using the annotated landmarks. However, this approach has many disadvantages and requires another evaluation scheme to confirm its medical credibility. Finally, domain adaptation is essential to transfer the proposed method to medical practice. Otherwise, introducing new acquisition equipment may result in an incorrect registration. This is important since even the model trained on the volumes acquired during the surgery cannot be applied to register the volumes acquired after the surgery.

We think that the segmentation-based methods are also interesting. We plan to verify the possibilities for automatic segmentation of the corresponding structures to make the method semi-supervised. This approach could result in higher generalizability and more stable training.

To conclude, we propose an algorithm dedicated to the registration of intraoperative US-US volumes. The proposed method does not require the similarity measure to be defined, is resistant to the resection process, and calculates the transformation in real-time. It may be useful for real-time accurate image-guidance during the surgery.

Acknowledgments. This work was funded by NCN Preludium project no. UMO-2018/29/N/ST6/00143 and NCN Etiuda project no. UMO-2019/32/T/ST6/00065. The authors declare no conflict of interest.

References

1. Holland, E.: Progenitor cells and glioma formation. Curr. Opin. Neurol. **14**(6), 683–688 (2001)
2. Ostrom, Q., et al.: CBTRUS statistical report: primary brain and other central nervous system tumors diagnosed in the United States in 2012–2016. Neuro Oncol. **21**, 1–100 (2019)
3. Schomas, D., et al.: Intracranial low-grade gliomas in adults: 30-Year experience with long-term follow-up at Mayo Clinic. Neuro Oncol. **11**(4), 437–445 (2009)
4. Jakola, A., et al.: Comparison of a strategy favoring early surgical resection vs a strategy favoring watchful waiting in low-grade gliomas. JAMA J. Am. Med. Assoc. **308**(18), 1881–1888 (2012)
5. Banerjee, J., Klink, C., Peters, E., Niessen, W., Moelker, A., van Walsum, T.: Fast and robust 3D ultrasound registration - block and game theoretic matching. Med. Image Anal. **20**(1), 173–183 (2015)
6. Che, C., Mathai, T., Galeotti, J.: Ultrasound registration: a review. Methods **115**, 128–143 (2017)
7. Wein, W., Ladikos, A., Fuerst, B., Shah, A., Sharma, K., Navab, N.: Global registration of ultrasound to MRI using the LC^2 metric for enabling neurosurgical guidance. In: Mori, K., Sakuma, I., Sato, Y., Barillot, C., Navab, N. (eds.) MICCAI 2013. LNCS, vol. 8149, pp. 34–41. Springer, Heidelberg (2013). https://doi.org/10.1007/978-3-642-40811-3_5
8. Xiao, Y., Fortin, M., Unsgärd, G., Rivaz, H., Reinertsen, I.: REtroSpective Evaluation of Cerebral Tumors (RESECT): a clinical database of pre-operative MRI and intra-operative ultrasound in low-grade glioma surgeries. Med. Phys. **44**(7), 3875–3882 (2017)
9. Drobny, D., Ranzini, M., Ourselin, S., Vercauteren, T., Modat, M.: Landmark-based evaluation of a block-matching registration framework on the RESECT pre- and intra-operative brain image data set. In: Zhou, L., et al. (eds.) LABELS/HAL-MICCAI/CuRIOUS 2019. LNCS, vol. 11851, pp. 136–144. Springer, Cham (2019). https://doi.org/10.1007/978-3-030-33642-4_15
10. Luo, J., et al.: A feature-driven active framework for ultrasound-based brain shift compensation. In: Frangi, A.F., Schnabel, J.A., Davatzikos, C., Alberola-López, C., Fichtinger, G. (eds.) MICCAI 2018. LNCS, vol. 11073, pp. 30–38. Springer, Cham (2018). https://doi.org/10.1007/978-3-030-00937-3_4
11. Machado, I., et al.: Non-rigid registration of 3D ultrasound for neurosurgery using automatic feature detection and matching. Int. J. Comput. Assist. Radiol. Surg. **13**(10), 1525–1538 (2018). https://doi.org/10.1007/s11548-018-1786-7
12. Canalini, L., Klein, J., Miller, D., Kikinis, R.: Registration of ultrasound volumes based on euclidean distance transform. In: Zhou, L., Reinertsen, I. (eds.) LABELS/HAL-MICCAI/CuRIOUS 2019. LNCS, vol. 11851, pp. 127–135. Springer, Cham (2019). https://doi.org/10.1007/978-3-030-33642-4_14
13. Fan, J., Cao, X., Wang, Q., Yap, P., Shen, D.: Adversarial learning for mono- or multi-modal registration. Med. Image Anal. **58**, 101545 (2019)
14. Mahapatra, D., Antony, B., Sedai, S., Garnavi, R.: Deformable medical image registration using generative adversarial networks. In: International Symposium on Biomedical Imaging (ISBI), pp. 1449–1453 (2018)
15. Balakrishnan, G., Zhao, A., Sabuncu, M., Guttag, J., Dalca, A.: VoxelMorph: a learning framework for deformable medical image registration. IEEE Trans. Med. Imaging **38**(8), 1788–1800 (2019)

16. Haskins, G., Kruger, U., Yan, P.: Deep learning in medical image registration: a survey. Mach. Vis. Appl. **31**(1), 1–18 (2020)
17. Xu, Z., et al.: Evaluation of six registration methods for the human abdomen on clinically acquired CT. IEEE Trans. Biomed. Eng. **63**(8), 1563–1572 (2016)
18. Dalca, A., Hering, A., Hansen, L., Heinrich, M.: The Learn2Reg Challenge (2020). https://learn2reg.grand-challenge.org

Robust Ultrasound-to-Ultrasound Registration for Intra-operative Brain Shift Correction with a Siamese Neural Network

Amir Pirhadi[1(✉)], Hassan Rivaz[1], M. Omair Ahmad[1], and Yiming Xiao[2]

[1] Department of Electrical and Computer Engineering, Concordia University,
Montréal, QC, Canada
a_pirhad@encs.concordia.ca

[2] Department of Computer Science and Software Engineering, Concordia University,
Montréal, QC, Canada

Abstract. In brain tumor resection, soft tissue shift (called brain shift) can displace the surgical target and render the surgical plan invalid. Intra-operative ultrasound (iUS) with robust image registration algorithms can effectively correct brain shift to ensure quality of resection and patient safety. Herein, we proposed a novel technique to automatically align iUS scans acquired before and after tumor resection, in order to confirm removal of cancerous tissues while minimizing resection of healthy tissue. More specifically, we employed a Siamese network to locate matching anatomical landmarks within iUS scans. Selected landmarks were used to search for the best affine transformation to align iUS obtained at different surgical stages. The proposed method was validated with the publicly available REtroSpective Evaluation of Cerebral Tumors (RESECT) database. After image alignment, the mean target registration error (mTRE) was effectively reduced from $3.55 \pm 1.76\,\mathrm{mm}$ to $1.26 \pm 0.48\,\mathrm{mm}$ in before and after resection and from $3.49 \pm 1.56\,\mathrm{mm}$ to $1.16 \pm 0.49\,\mathrm{mm}$ in before and during resection. In general, the results are comparable to the state-of-the-art techniques, validated on the same database, and our technique demonstrated excellent performance in iUS-based brain shift correction for optimal therapeutic outcomes.

1 Introduction

In brain tumor surgery, soft tissue deformation, or brain shift, can result from many factors, such as gravity and drug administration, and can greatly affect the quality and safety of the procedure. Intra-operative imaging is often used to track brain shift and the surgical progress. In contrast to the high cost and special setups required by intra-operative magnetic resonance imaging (iMRI), intra-operative ultrasound (iUS) is a cost-effective and portable imaging modality that has gained popularity in the clinic [1]. However, to help account for brain shift to update pre-surgical plans [2] in commonly used surgical navigation system,

© Springer Nature Switzerland AG 2021
J. A. Noble et al. (Eds.): ASMUS 2021, LNCS 12967, pp. 85–95, 2021.
https://doi.org/10.1007/978-3-030-87583-1_9

robust and efficient image registration algorithms are crucial. While so far most of the previous works [3–5] focus on the alignment of pre-operative MRI and iUS obtained before dura-opening, very few have attempted to correct additional tissue deformation during the procedure, which is also important to ensure clean removal of any residual tumour and thus increase the patient survival rate [6]. In this scenario, iUS-iUS registration is required, and poses unique challenges from the more commonly seen MRI-iUS alignment. For example, in addition to continuous tissue deformation introduced from gravity and tissues removal, the procedure can also significantly alter image features and reduce image quality in iUS by introducing air bubbles, debris, and blood clots in the surgical site, rendering registration of pre- and post-resection iUS images more challenging.

To tackle the discrepancies of iUS at different surgical stages, an attractive solution is based on matching anatomical landmarks that are consistent between scans [7]. To date, a number of medical image registration algorithms based on automatic landmark detection have been proposed. Lu *et al.* [8] and Urschler *et al.* [9] used segmentation and corner detection for finding the global shape information. Local keypoint selection and feature matching were done using scale-invariant feature transform (SIFT). They applied their method on 2D US images of kidney and thoracic 3D CT images respectively in the application of feature-based non-rigid registration. Machado *et al.* [10] presented an optimal global feature mapping in 3D iUS images in neurosurgery using 3D SIFT-Rank, which showed a large improvement in the alignment. However, the fully automatic aspect of their work can potentially make it sensitive to selection of voxels in and around the tumor, which usually do not match the post-resection scan.

Application of deep learning in medical image registration has rapidly increased during the past few years. Canalini *et al.* [11] proposed segmentation-based registration of 3D iUS images in neurosurgery. They segmented hyper-echogenic regions of the brain that keep their correspondence after resection and excluded the resection cavity. An attractive alternative to this approach is to exploit Siamese networks, which require substantially less training data and have shown promising results in tracking tasks in the computer vision field [12–14]. Gomariz *et al.* [15] took advantage of this strength to develop a Siamese network for tracking 2D US images of the liver. Since landmark locations are not expected to change drastically between frames, they utilized a temporal consistency model to weight the similarity map around the previous landmark location.

In this article, we proposed a novel technique based on Siamese networks for landmark tracking in the 3D iUS images. This network was chosen for two main reasons. First, it is an end-to-end learning technique. Second, it works on new domains not seen by the network in the training stage as the network extracts general features from the inputs that are necessary for comparison [16]. To allow interactivity and flexibility, in our method, template landmarks are first manually selected by clinicians in pre-resection images. Then, with the Siamese network, matching landmarks can be quickly identified after resection starts to continuously track brain shift. Our main contributions are listed below:

1. Using a Siamese network in the application of landmark tracking for iUS-based brain shift correction at different surgical stages (during and after resection).
2. Demonstrating the adaptation of the Siamese network from natural images to US volumes without re-training. This suggests its high adaptability to scans from different machines, imaging settings, and anatomies.
3. Employing a 2.5D search scheme for efficient and robust 3D landmark tracking.
4. Fast registration with an iterative re-weighted least squares (IRLS) algorithms to ensure robustness, rendering the method an attractive choice in neurosurgery.

2   Methods

To automatically match clinician-defined reference landmarks from pre-resection iUS scans in those after resection starts, we are inspired by object tracking in videos with Siamese neural networks [14], where high-dimensional image features are represented robustly in more efficient form with convolutional neural networks (CNNs). With the automatically identified matching landmark pairs, an affine image alignment is then estimated. Here, we denote the pre-, during- and post-resection US images as iUS_{pre}, iUS_{during} and iUS_{post}, respectively.

2.1   Siamese Network

An overview of our fully convolutional Siamese network is shown in Fig. 1. In essence, the network finds an embedding function Φ to extract a representative feature map of the input image. The embedded images then would be passed through a cross-correlation layer as a similarity function, to find the location of the template image inside the search image. It has been shown that this implementation is fully-convolutional based on the search image. That means translation is commutative as shown below.

$$\Phi(L_\tau x) = L_\tau \Phi(x) \tag{1}$$

where L is the translation function with translation τ and x is the search image. The fully-convolutional nature of the network enables us to use a large search image. The search image is divided into sub-windows. These sub-windows are passed through the network. Similarity of the embedded template image and all the translated sub-windows of search image would be evaluated in the cross-correlation layer at once. In our method, we used the positions of the pre-selected landmarks in iUS_{pre} to find their correspondences in iUS_{during} and iUS_{post}.

2.2 Training

The network was trained on the ILSVRC17 dataset [18] for video object tracking purposes [19], and no network fine-tuning was performed for our ultrasound application. Stochastic gradient decent (SGD) with binary cross-entropy loss was used, and the learning rate was set to 0.01 with a batch size of 8. The template image size is 127×127 and the search image size is 255×255. For optimization, a binary ground truth (match vs. not a match) was generated by considering a radius of 4 pixels around the center of the similarity map. If the maximum similarity occurs within the radius, it indicates a match. Otherwise, it is not.

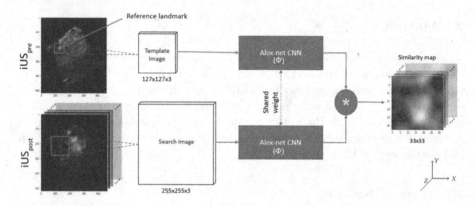

Fig. 1. Siamese network. The convolutional stage of AlexNet [17] was chosen for feature embedding thanks to its memory efficiency and good performance, and cross-correlation is used as the similarity function. Here, the x-y plane is used for demonstration.

2.3 Preprocessing

We used the RESECT database [20] to demonstrate our technique. Voxel sizes in the dataset differ among patients and even scans. As such, all US images were resampled to the smallest isotropic resolution in the database, i.e., $0.14 \times 0.14 \times 0.14 \, \text{mm}^3$. In iUS_{pre}, pre-identified anatomical landmarks will be used as references to find the corresponding ones in iUS_{during} or iUS_{post}. In the case of iUS_{pre} to iUS_{post} registration, iUS_{pre} and iUS_{post} will serve as the template and search images, respectively, and the images were cropped around the location of the reference landmark according to the requirement of the network. Finally, all image intensity ranges were normalized to $[0, 1]$.

2.4   The 2.5D Approach

The network we used in this paper obtains 2D images as inputs. However, our goal is to find landmarks in 3D US images. Similar to the approach of Heinrich *et al.* [21], we performed landmark matching in three orthogonal directions, or 2.5D. In each direction, a region of interest (ROI) around a reference landmark in iUS_{pre} and a larger one in iUS_{post}/iUS_{during} were selected as template and search image respectively. To enable 3D search, the ROI in iUS_{post}/iUS_{during} was swept forward and backward in that direction with a stride of 2, so the search images have one slice overlap.

The location of the maximum correlation in the similarity map was considered as the desired result. Since we had three searching directions by using the 2.5D approach, at the end, we have 3 predictions for each landmark's location in iUS_{post} or iUS_{during}. A final decision was made base on them. If at least two of these three predicted landmarks were located near each other ($<2\,\mathrm{mm}$), the average of them would be considered as the result. Otherwise, the results were treated as incorrect and were discarded.

2.5   Affine Transformation

Affine transformation has been utilized for tissue shift correction in brain tumour resection thanks to its ability to robustly improve global misalignment. Furthermore, it is simpler and faster in comparison to more complex deformation models, such as free-form B-splines [5]. To estimate a 12-parameter 3D affine transformation, at least 4 pairs of landmarks are required to solve a linear system. Our landmark selection method usually provides at least 5 landmarks (Tables 1 and 2), resulting in an over-determined linear system that can be solved.

In order to obtain the optimal 3D affine transformation while overcoming the potential influence of outlier landmarks, we employed the iterative re-weighted least square (IRLS) method [22]. Here, the Cauchy function (Eq. 2) has been chosen as the weighing function in the IRLS algorithm, where small weights were assigned to the outliers in the linear equation [23] to mitigate their impacts. In Eq. 2, r_i is the residual after an iteration and $R = 1$ was selected manually.

$$w\left(r_i\right) = \frac{1}{1 + \left(r_i/R\right)^2} \tag{2}$$

Table 1. mTREs of our method and two comparison methods [10,11] for iUS_{pre} vs. iUS_{post} registration. Initial mTRE before registration and minimum achievable mTRE (with affine transformations) were calculated from the ground truth landmarks. Note that nonlinear registration results can be lower than min achievable mTREs.

Patient ID	No. total landmarks	No. selected landmarks	Initial mTRE (mm)	After Siamese based affine (mm)	Canalini et al. (mm)	Machado et al. (mm)	Affine Minimum achievable mTRE (mm)
1	13	10	5.80 (3.62–7.22)	1.32 (0.47–4.06)	1.03	1.48	0.97
2	10	5	3.65 (1.71–6.72)	**2.47** (0.45–4.35)	3.90	2.62	1.57
3	11	6	2.91 (1.53–4.30)	1.05 (0.28–1.79)	1.15	1.04	0.67
4	12	9	2.22 (1.25–2.94)	0.81 (0.21–1.43)	0.61	0.83	0.55
6	11	6	2.12 (0.75–3.82)	1.66 (0.29–3.72)	1.41	1.55	1.20
7	18	15	3.62 (1.19–5.93)	**1.73** (0.41–4.39)	2.03	2.38	1.50
12	11	9	3.97 (2.58–6.35)	1.36 (0.25–3.42)	0.79	1.20	0.98
14	17	17	0.63 (0.17–1.76)	0.53 (0.28–0.90)	0.46	0.53	0.45
15	15	15	1.63 (0.62–2.69)	0.72 (0.25–1.84)	0.58	0.74	0.70
16	17	13	3.13 (0.82–5.41)	1.19 (0.42–2.92)	0.92	1.94	1.08
17	11	9	5.71 (4.25–8.03)	1.40 (0.49–4.20)	1.10	1.99	0.96
18	13	6	5.29 (2.94–9.26)	1.29 (0.45–2.90)	1.13	1.69	1.14
19	13	11	2.05 (0.43–3.24)	1.23 (0.32–5.28)	1.10	2.78	0.86
21	9	7	3.35 (2.34–5.64)	1.77 (0.76–3.71)	1.80	1.07	0.76
24	14	11	2.61 (1.96–3.41)	1.02 (0.35–2.54)	0.87	1.35	0.62
25	12	11	7.61 (6.40–10.25)	**1.20** (0.29–2.42)	1.21	1.24	0.91
27	12	11	3.98 (3.09–4.82)	0.63 (0.20–1.06)	0.53	0.83	0.47
Mean	13	10	3.55	1.26	1.21	1.49	0.90
Std			1.76	0.48	0.81	0.67	0.33

2.6 Experimental Setup

We validated our registration method using 17 clinical cases that have pre-, during- and post-resection iUS images in the RESECT public database [20], where matching ground truth landmark pairs have been provided by experts. Quantitative evaluation for our algorithm was performed using mean target registration errors (mTREs) before and after registration using the ground truth landmark pairs. The metric is shown in Eq. 3.

$$mTRE = \frac{1}{N} \sum_{i=1}^{N} \| T(x_i) - x_i' \|, \tag{3}$$

where x_i and x_i' are the landmark pairs in the corresponding iUS scans, T is the affine transformation estimated with the proposed method, and N is the total number of landmarks. Here, we used the full set of landmark pairs from the original database to compute mTRE. The accuracy of our method was compared against two recent nonlinear techniques [10,11] that were validated on the same database. Statistical tests were done to confirm the performance of our method.

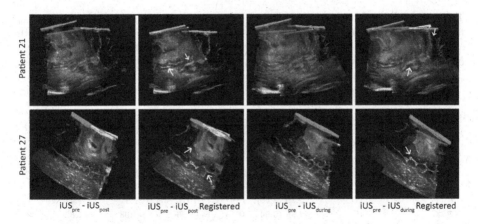

Fig. 2. Visual comparison between image pairs before and after registration with our proposed method. Cyan color $= iUS_{post}$ and iUS_{during} and orange color $= iUS_{pre}$. The arrows mark the sulci regions of improved alignment.

3   Results

Tables 1 and 2 show the quantitative evaluation of mTREs, with the number of landmarks selected by our method to obtain the affine transformations. On average, the tested cases in the RESECT database have 13 and 21 landmarks per patient in $iUS_{pre} - iUS_{post}$ and $iUS_{pre} - iUS_{during}$ cases, respectively. With automatic landmark selection using the Siamese network, we can obtain an average of 10 and 18 landmarks per patient, respectively. With our registration method, the initial misalignment of 3.55 ± 1.76 mm was reduced to 1.26 ± 0.48 mm between before and after resection, and from 3.49 ± 1.56 mm to 1.16 ± 0.49 mm between before and during the resection. Furthermore, the comparison between our method and those by Machado *et al.* [10] and Canalini *et al.* [11] are shown in Tables 1 and 2. Overall, our proposed approach showed very similar performance to that of Canalini *et al.* [24] and on average outperformed the method of Machado *et al.* [10] in iUS_{pre} vs. iUS_{post} registration. In addition, our results are also better in iUS_{pre} vs. iUS_{during} registration than Canalini *et al.* [11] on average (Machado *et al.* [10] didn't perform iUS_{pre} vs. iUS_{during} registration).

Table 2. mTREs of our method and a comparison method [11] for iUS_{pre} vs. iUS_{during} registration. Initial mTRE before registration and minimum achievable mTRE (with affine transformations) are calculated based on the ground truth landmarks provided. Note that nonlinear registration results can be lower than min achievable mTREs.

Patient ID	No. total landmarks	No. selected landmarks	Initial mTRE (mm)	After Siamese based affine (mm)	Canalini et al. (mm)	Affine Minimum achievable mTRE (mm)
1	34	33	2.32 (1.49–3.29)	0.88 (0.18–1.90)	0.64	0.83
2	16	11	3.10 (1.79–5.19)	**1.43** (0.33–4.42)	1.50	1.21
3	17	17	1.93 (0.67–3.02)	0.79 (0.26–1.33)	0.77	0.70
4	19	17	4.00 (3.03–5.22)	0.89 (0.30–2.44)	0.80	0.74
6	21	17	5.19 (2.60–7.18)	**1.77** (0.32–3.42)	5.17	1.47
7	22	19	4.69 (0.94–8.16)	2.46 (0.26–6.56)	1.98	1.82
12	24	23	3.39 (1.74–4.81)	1.12 (0.15–2.01)	0.84	1.04
14	22	22	0.71 (0.42–1.59)	0.50 (0.03–0.90)	0.41	0.47
15	21	21	2.04 (0.85–2.84)	0.68 (0.27–1.38)	0.60	0.58
16	19	10	3.19 (1.22–4.53)	1.51 (0.21–4.60)	1.26	1.10
17	17	11	6.32 (4.65–8.07)	1.54 (0.49–3.91)	1.49	0.97
18	23	16	5.06 (1.55–7.44)	1.33 (0.20–3.71)	1.18	1.10
19	21	20	2.06 (0.42–3.40)	1.03 (0.20–2.35)	0.96	0.90
21	18	14	5.10 (3.37–5.94)	1.25 (0.33–3.81)	1.11	0.97
24	21	19	1.76 (1.16–2.65)	0.78 (0.13–1.85)	0.67	0.64
25	20	19	3.60 (2.19–5.02)	0.72 (0.27–2.01)	0.55	0.65
27	16	16	4.93 (3.61–7.01)	0.96 (0.19–2.34)	0.87	0.57
Mean	20	18	3.49	1.16	1.22	0.93
Std			1.56	0.49	1.10	0.35

We performed Wilcoxon rank sum tests on the mTREs before and after registration with the proposed method, as well as on the same metric to compare between our method and those two recent works [10,11]. The statistical tests showed that the reduction in mTREs with our technique was statistically significant ($p < 0.001$). In addition, in post-resection registration, our results are comparable (p > 0.05) to those by Machado *et al.* and Canalini *et al.* [10,11] while our average mTRE reduction is better than that of Machado *et al.* [10]. In during-resection registration, our results were comparable to those of Cananili *et al.* [11] ($p = 0.19$), with a better average mTRE. Furthermore, qualitative assessment of our method is illustrated in Fig. 2, where pre-resection and during-/post-resection images are overlaid in cases of before and after registration for two patients. Note that anatomical features (e.g., sulci) are shown as hyperintense edges in each image.

4 Discussion

In this paper, we used the 12-parameter affine transformation for brain shift correction during tumour resection. Although nonlinear deformation models, such as B-splines, can more precisely adapt to local tissue deformation, the computational complexity is much higher. In addition, robustness and reliability in intra-operative registration algorithms can be more valuable in the clinic, and thus affine transformation appears more advantageous, considering its mTRE measures are comparable to the non-linear counterparts [10,11].

In the proposed algorithm, reference landmarks for tracking need to be identified first while in previous fiducial point-based registration methods, automatic landmark selections were employed based on image feature detection (e.g., SIFT). The involvement of manual interaction with the image can help ensure the distribution of salient anatomical landmarks for optimal registration quality [25], and improve the flexibility and robustness in real clinical applications. The proposed method showed an excellent performance in reducing the initial misalignment in terms of mTREs. On average, the results are comparable to or better than the recent state-of-the-art techniques [10,11], which were validated also on the RESECT database. For our method, the whole process of landmark matching and registration on NVIDIA GeForce GTX 1050 Ti GPU took about 45 s in post-resection registration and 70 s in during-resection registration (due to a greater number of landmarks). The segmentation-based registration by Canalini et al. [11] takes 55 s on average and the SIFT-based registration by Machado et al. [10] is the fastest with 30 s of average run time. Thus, the computational time of our technique is highly promising in real clinical applications and comparable to the state-of-the-art methods.

One limitation of our proposed method lies in the requirement of reference landmark tagging, which can cost extra clinical time. However, as with affine transformation, the number of landmarks doesn't need to be large, and with the rich experience of the clinicians, this can be performed robustly and quickly, especially considering that a typical neurosurgery lasts a few hours. This also offers more flexibility for manual interaction. Although we present the proposed technique for iUS-iUS registration, we believe that it can be adapted to MR-iUS registration, which will be investigated in the future.

5   Conclusion

We have proposed a robust and efficient iUS-iUS registration technique based on anatomical landmark detection to account for tissue shift in brain tumor surgery. The method demonstrated excellent performance compared to the recent works, and can potentially improve the accuracy and safety of the procedure. The Siamese network weights were trained on natural images without domain-specific fine-tuning, rendering the method robust to scanner differences.

References

1. Unsgaard, G., et al.: Intra-operative 3d ultrasound in neurosurgery. Acta Neurochir. **148**(3), 235–253 (2006)
2. Xiao, Y., Eikenes, L., Reinertsen, I., Rivaz, H.: Nonlinear deformation of tractography in ultrasound-guided low-grade gliomas resection. Int. J. Comput. Assist. Radiol. Surg. **13**(3), 457–467 (2018)
3. Wein, W.: Brain-shift correction with image-based registration and landmark accuracy evaluation. In: Stoyanov, D., et al. (eds.) POCUS/BIVPCS/CuRIOUS/CPM 2018. LNCS, vol. 11042, pp. 146–151. Springer, Cham (2018). https://doi.org/10. 1007/978-3-030-01045-4_17
4. Heinrich, M.P.: Intra-operative ultrasound to MRI fusion with a public multimodal discrete registration tool. In: Stoyanov, D., et al. (eds.) POCUS/BIVPCS/CuRIOUS/CPM 2018. LNCS, vol. 11042, pp. 159–164. Springer, Cham (2018). https://doi.org/10.1007/978-3-030-01045-4_19
5. Masoumi, N., Xiao, Y., Rivaz, H.: ARENA: inter-modality affine registration using evolutionary strategy. Int. J. Comput. Assist. Radiol. Surg. **14**(3), 441–450 (2019)
6. Marko, N.F., Weil, R.J., Schroeder, J.L., Lang, F.F., Suki, D., Sawaya, R.E.: Extent of resection of glioblastoma revisited: personalized survival modeling facilitates more accurate survival prediction and supports a maximum-safe-resection approach to surgery. J. Clin. Oncol. **32**(8), 774 (2014)
7. Xiao, Y., et al.: Evaluation of MRI to ultrasound registration methods for brain shift correction: the curious2018 challenge. IEEE Trans. Med. Imaging **39**(3), 777–786 (2020)
8. Lu, X., Zhang, S., Yang, W., Chen, Y.: Sift and shape information incorporated into fluid model for non-rigid registration of ultrasound images. Comput. Methods Programs Biomed. **100**(2), 123–131 (2010)
9. Urschler, M., Bauer, J., Ditt, H., Bischof, H.: SIFT and shape context for feature-based nonlinear registration of thoracic CT images. In: Beichel, R.R., Sonka, M. (eds.) CVAMIA 2006. LNCS, vol. 4241, pp. 73–84. Springer, Heidelberg (2006). https://doi.org/10.1007/11889762_7
10. Machado, I., et al.: Non-rigid registration of 3d ultrasound for neurosurgery using automatic feature detection and matching. Int. J. Comput. Assist. Radiol. Surg. **13**(10), 1525–1538 (2018)
11. Canalini, L., Klein, J., Miller, D., Kikinis, R.: Enhanced registration of ultrasound volumes by segmentation of resection cavity in neurosurgical procedures. Int. J. Comput. Assist. Radiol. Surg. **15**(12), 1963–1974 (2020). https://doi.org/10.1007/s11548-020-02273-1
12. Guo, Q., Feng, W., Zhou, C., Huang, R., Wan, L., Wang, S.: Learning dynamic Siamese network for visual object tracking. In: Proceedings of the IEEE International Conference on Computer Vision, pp. 1763–1771 (2017)
13. He, A., Luo, C., Tian, X., Zeng, W.: A twofold Siamese network for real-time object tracking. In: Proceedings of the IEEE Conference on Computer Vision and Pattern Recognition, pp. 4834–4843 (2018)
14. Bertinetto, L., Valmadre, J., Henriques, J.F., Vedaldi, A., Torr, P.H.S.: Fully-convolutional Siamese networks for object tracking. In: Hua, G., Jégou, H. (eds.) ECCV 2016. LNCS, vol. 9914, pp. 850–865. Springer, Cham (2016). https://doi. org/10.1007/978-3-319-48881-3_56

15. Gomariz, A., Li, W., Ozkan, E., Tanner, C., Goksel, O.: Siamese networks with location prior for landmark tracking in liver ultrasound sequences. In: 2019 IEEE 16th International Symposium on Biomedical Imaging (ISBI 2019), pp. 1757–1760. IEEE (2019)

16. Koch, G., Zemel, R., Salakhutdinov, R.: Siamese neural networks for one-shot image recognition. In: ICML Deep Learning Workshop, vol. 2. Lille (2015)

17. Krizhevsky, A., Sutskever, I., Hinton, G.E.: ImageNet classification with deep convolutional neural networks. Adv. Neural. Inf. Process. Syst. **25**, 1097–1105 (2012)

18. Russakovsky, O., et al.: ImageNet large scale visual recognition challenge. Int. J. Comput. Vision **115**(3), 211–252 (2015)

19. Pytorch-SiamFC. https://github.com/rafellerc/Pytorch-SiamFC. Accessed 29 June 2021

20. Xiao, Y., Fortin, M., Unsgård, G., Rivaz, H., Reinertsen, I.: Retrospective evaluation of cerebral tumors (RESECT): a clinical database of pre-operative MRI and intra-operative ultrasound in low-grade glioma surgeries. Med. Phys. **44**(7), 3875–3882 (2017)

21. Heinrich, M.P., Hansen, L.: Highly accurate and memory efficient unsupervised learning-based discrete CT registration using 2.5D displacement search. In: Martel, A.L., et al. (eds.) MICCAI 2020. LNCS, vol. 12263, pp. 190–200. Springer, Cham (2020). https://doi.org/10.1007/978-3-030-59716-0_19

22. Holland, P.W., Welsch, R.E.: Robust regression using iteratively reweighted least-squares. Commun. Stat. Theory Methods **6**(9), 813–827 (1977)

23. Rivaz, H., Boctor, E.M., Choti, M.A., Hager, G.D.: Real-time regularized ultrasound elastography. IEEE Trans. Med. Imaging **30**(4), 928–945 (2010)

24. Canalini, L., Klein, J., Miller, D., Kikinis, R.: Segmentation-based registration of ultrasound volumes for glioma resection in image-guided neurosurgery. Int. J. Comput. Assist. Radiol. Surg. **14**(10), 1697–1713 (2019). https://doi.org/10.1007/s11548-019-02045-6

25. Luo, J., et al.: Do public datasets assure unbiased comparisons for registration evaluation? arXiv preprint arXiv:2003.09483 (2020)

Pose Estimation of 2D Ultrasound Probe from Ultrasound Image Sequences Using CNN and RNN

Kanta Miura[1]([✉]), Koichi Ito[1], Takafumi Aoki[1], Jun Ohmiya[2], and Satoshi Kondo[2]

[1] Graduate School of Information Sciences, Tohoku University, 6-6-05, Aramaki Aza Aoba, Aoba-ku, Sendai-shi, Miyagi 9808579, Japan
kanta@aoki.ecei.tohoku.ac.jp
[2] AI Technology Development Division IoT Service Platform Development Operations, Konica Minolta, Inc., 1-2, Sakura-machi, Takatsuki-shi, Osaka 5698503, Japan

Abstract. In this paper, we propose an ultrasound (US) probe pose estimation method only from US image sequences using deep learning for volume reconstruction. The proposed method employs the combination of convolutional neural network (CNN) and recurrent neural network (RNN) to estimate the US probe pose in light of the long-term temporal information of US image sequences. The features extracted by CNN are input to RNN to estimate the relative and absolute pose of the US probe. Through a set of experiments using US image sequence datasets with ground-truth pose measured by an optical tracking system, we demonstrate that the proposed method exhibits the efficient performance on US probe pose estimation and volume reconstruction compared with the conventional method.

Keywords: Ultrasound · Volume reconstruction · RNN · CNN · Probe pose estimation

1 Introduction

Medical volume data is a visualization of the three-dimensional (3D) structure of the body, which is essential for detecting diseases and determining medical treatment strategies. CT and MRI are commonly used as major medical volume data, while 3D ultrasound (US) images reconstructed from US images have been attracting much attention in recent years because of the advantages of US. US imaging has the advantages of high spatial resolution, real-time imaging, and non-invasiveness, and is less burdensome to the subjects than CT and MRI. In addition, US systems have become smaller and less expensive, making them

This work was supported in part by the WISE Program for AI Electronics, Tohoku University.

J. A. Noble et al. (Eds.): ASMUS 2021, LNCS 12967, pp. 96–105, 2021.
https://doi.org/10.1007/978-3-030-87583-1_10

easier to acquire US images. If we can analyze blood vessels and muscles in 3D using volume data reconstructed only from US images, we can analyze the inside of the body without choosing a location since US images can be acquired using only a 2D US probe. Therefore, 3D US images are extremely useful for point-of-care testing and sports medicine. For this purpose, it is necessary to obtain the probe pose with high accuracy.

Many methods have been proposed to measure the pose of the US probe using an external device such as an electromagnetic device, an optical device, and a camera [4,6–8,10,15,16,18,19]. These methods require an external device to be attached to the 2D US probe, which sacrifices the smooth scanning that is an advantage of US imaging and also increases the cost.

To address this problem, methods of estimating the probe pose using convolutional neural network (CNN) only from US image sequences have been proposed [5,11,12,14]. Prevost et al. [14] proposed CNN designed based on the speckle decorrelation model [3] to estimate the relative pose between two consecutive frames. In this method, the relative pose between each frame pair is estimated using only two US images, so the estimation error accumulates and the estimation accuracy becomes low. Their latest work in [13] improves on this method by using an inertial measurement unit (IMU), however IMU has to be attached to the probe. Miura et al. [11,12] used two CNNs: one to estimate optical flow and the other to extract features from US image sequences, based on the work of Prevost et al. [14] and proposed loss functions considering geometric consistency. Guo et al. [5] used 3D CNN with multiple consecutive US images as input instead of only two US images. The above methods only deal with the short-term temporal information of US image sequences, and it is difficult to consider the long-term temporal information.

In this paper, we propose a new 2D US probe pose estimation method only from US image sequences using deep learning. The proposed method employs recurrent neural network (RNN) in addition to CNN to estimate the probe pose based on the long-term temporal information. Inspired by Xue et al.'s approach [20] of combining CNN and RNN to estimate camera pose from stereo image sequences, the proposed method introduces a new network architecture which estimates the absolute pose in addition to the relative pose to reduce the accumulated errors. In addition, a large-scale dataset of US image sequences with the ground-truth probe position was created to evaluate the estimation accuracy of the proposed method. This dataset contains 410 scans acquired from the subject's forearm, a breast phantom, and a hypogastric phantom. The contributions of this paper are summarized below:

1. Propose a new CNN architecture that can estimate both relative and absolute pose by combining CNN and RNN,
2. Create a large-scale dataset of US image sequences with the ground-truth probe position, and
3. Demonstrate the effectiveness of the proposed method in pose estimation of US probe for 3D US volume reconstruction compared with the conventional method through experiments.

2   Methods

In this section, we describe the method proposed in this paper to estimate the pose of the US probe. In the conventional methods [11–14], two frames of the US image sequence are used as input to estimate the relative pose of the US probe. Since the probe moves continuously and over a long period of time, its motion should be estimated taking into account the long time information. Therefore, the proposed method uses RNN, which can aggregate features in the temporal direction, in addition to CNN, which extracts features from images between two frames. Both the relative pose $p_{i-1,i}$ between the frames $i-1$ and i and the absolute pose p_i of the frame i are estimated to reduce the accumulated error. $p_{i-1,i}$ and p_i consist of 6-parameter vectors: 3D rotation $(\theta_x, \theta_y, \theta_z)$ and 3D translation (t_x, t_y, t_z). Figure 1 shows the overview of the proposed method. First, the feature map between two consecutive frames is extracted using CNN. Next, the relative pose is estimated from the feature map using RNN. Then, we aggregate the feature map extracted by CNN, the hidden states of RNN that estimates the relative pose, and the output of RNN that estimates the absolute pose in the previous frame. Finally, the absolute pose of the current frame is estimated from the aggregated feature map using RNN. The details of each part are described below.

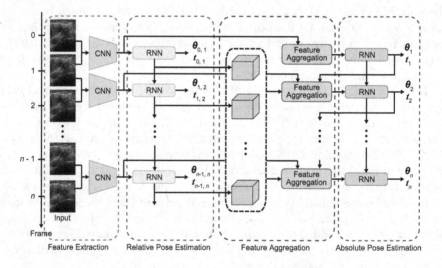

Fig. 1. Overview of the proposed method.

2.1   Feature Extraction

Figure 2(a) shows the detail of feature extraction. Since optical flow is useful in pose estimation demonstrated by the previous work [11–14], we also use an optical flow as an additional input. The optical flow is estimated by FlowNetS [1] trained on Flying Chairs dataset[1] and is concatenated with two US images along the channel direction. We employ AlexNet [9] trained on ImageNet provided by torchvision[2] as CNN to extract a feature map. Note that the global average pooling (GAP) layer and 3 fully-connected (FC) layers of AlexNet are not used. The output of AlexNet is used as the feature map and $X_i \in \mathbb{R}^{C \times H \times W}$ is the feature map between frames $i - 1$ and i, where C, H, and W represent the channel, height and width of the feature map, respectively.

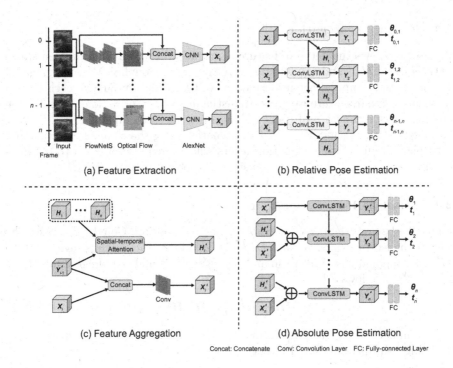

Fig. 2. Flow of each process in the proposed method: (a) feature extraction, (b) relative pose estimation, (c) feature aggregation, and (d) absolute pose estimation.

2.2 Relative Pose Estimation

Figure 2(b) shows the detail of relative pose estimation. The relative pose $p_{i-1,i}$ between frames $i - 1$ and i is estimated from the feature X_i. Convolutional LSTM (ConvLSTM) [17], which is a combination of convolution and LSTM, is used to utilize the temporal information of US image sequences as RNN. LSTM uses a 1D feature vector as input, while ConvLSTM uses a 2D feature map as input and can also use the spatial information. ConvLSTM takes the feature

---

[1] https://lmb.informatik.uni-freiburg.de/resources/datasets/.
[2] https://pytorch.org/vision/stable/.

map X_i and the hidden state at the previous frame H_{i-1} as input and outputs a feature map $Y_i \in \mathbb{R}^{C \times H \times W}$ and a hidden state $H_i \in \mathbb{R}^{C \times H \times W}$ as follows:

$$Y_i, H_i = \mathcal{U}(X_i, H_{i-1}), \tag{1}$$

where $\mathcal{U}$ represents ConvLSTM. Then, the relative pose is estimated from Y_i using two fully-connected layers. A set of hidden states $\{H_1, H_2, \cdots, H_n\}$ are aggregated into one feature map in the feature aggregation step described next.

2.3 Feature Aggregation

Figure 2(c) shows the detail of feature aggregation. First, a set of hidden states $\{H_1, H_2, \cdots, H_n\}$ and the feature map Y'_{i-1} from ConvLSTM are aggregated by the spatial-temporal attention module (STAM) [20]. In STAM, each hidden state and its channels are weighted to distinguish the information related to the absolute pose estimation of frame i, and a new feature map $H'_i \in \mathbb{R}^{C \times H \times W}$ is obtained. Next, X_i and Y'_{i-1} are concatenated along the channel direction and then pass through two convolution layers with kernel size of 3×3 to obtain a feature map $X'_i \in \mathbb{R}^{C \times H \times W}$, whose size corresponds to that of H'_i.

2.4 Absolute Pose Estimation

Figure 2(d) shows the detail of absolute pose estimation. ConvLSTM takes the sum of the feature maps X'_i and H'_i and the hidden state at the previous frame H''_{i-1} as input and a feature map $Y'_i \in \mathbb{R}^{C \times H \times W}$ and a hidden state $H''_i \in \mathbb{R}^{C \times H \times W}$ as follows:

$$Y'_i, H''_i = \mathcal{U}(X'_i + H'_i, H''_{i-1}). \tag{2}$$

Then, similar to relative pose estimation, the absolute pose is estimated from Y'_i using two fully-connected layers.

2.5 Loss Function

This section describes the loss functions used in training of the proposed method. As for relative pose estimation, we employ three loss functions used in [12]: Euclidean distance L_e, forward consistency loss L_f, and backward consistency loss L_b. For more details on these loss functions, see [12]. The combination of the three loss functions is used in relative pose estimation as follows:

$$L_{\text{relative}} = \alpha L_e + \beta L_f + \gamma L_b, \tag{3}$$

where α, β, and γ are the weights which balance each loss function. As for absolute pose estimation, we employ the loss function used in [20] as follows:

$$L_{\text{absolute}} = \sum_{i=1}^{N} \frac{1}{i} \|p_i - \hat{p}_i\|_2, \tag{4}$$

where $\boldsymbol{p}_i$ is the ground-truth pose, $\hat{\boldsymbol{p}}_i$ is the estimated absolute pose, and N is the number of input pairs. The overall loss function L_{all} for training the proposed method is given by

$$L_{\text{all}} = L_{\text{relative}} + L_{\text{absolute}}. \tag{5}$$

3   Materials

An US system and an optical tracking system are used to create a large-scale dataset of US image sequences with the scanning position of the US probe since no public dataset is available. We use SONIMAGE HS1 (Konica Minolta, Inc.) as the US system and OptiTrack V120: Trio (Acuity, Inc.) as the optical tracking system. A L18-4 linear probe (center frequency: 10 MHz) and a C5-2 convex probe (center frequency: 3.5 MHz) are used to acquire US images. 5 markers are attached on the probe to track its scanning position. The recording time is about 8 s at 30 fps, and the size of each US image is 442×526 pixels. The US system and the optical tracking system are synchronized during data acquisition. This dataset consists of 410 scans captured from the forearm of 6 subjects, a breast phantom, and a hypogastric phantom. The number of scans is 230 for the forearm, 140 for the breast phantom, and 40 for the hypogastric phantom, respectively.

4   Experiments

This section describes the experiments of evaluating the proposed method.

4.1   Experimental Condition

The proposed method is trained in two ways as follows. First, we use 50 scans acquired from the breast phantom with the probe moved in a straight line out of 410 scans to evaluate the basic performance of the proposed method. The 50 scans are divided into 40 scans for training, 5 scans for validation, and 5 scans for evaluation. Next, we use 410 scans acquired from the three types of scanning targets with various movements of the probe. The 410 scans are divided into 300 scans for training, 30 scans for validation, and 80 scans for evaluation.

Sharpness-aware minimization (SAM) [2] is used as an optimizer and the learning rate is set to 5e−5. Note that we use Adam as the base optimizer of SAM and set the weight decay for preventing overfitting to 0.1. The batch size is 16, the number of epochs is 50, the number of input pairs N is 10, and 25% dropout is added after the fully-connected layers except the last one. US images with 442×526 pixels are cropped at 442×442 pixels in the center and then are resized to 128×128 pixels. The pixel value of each resized image is normalized to have zero mean and the unit variance. The weights of the loss functions are set to $\alpha = 10$, $\beta = 0.1$, and $\gamma = 0.1$. All the methods are implemented using Pytorch 1.7.1 on Intel(R) Xeon(R) W-2133 CPU 3.60 GHz with GeForce RTX

2080 Ti. We evaluate the accuracy of the estimated parameters by the sum of absolute errors of the relative pose parameters for each frame by

$$\mathrm{AE} = \sum_{i=1}^{M} |\boldsymbol{p}_{i-1,i} - \hat{\boldsymbol{p}}_{i-1,i}|, \tag{6}$$

Table 1. Summary of experimental results using the breast phantom.

Method	AE (degree/mm)						Final drift (mm)
	θ_x	θ_y	θ_z	t_x	t_y	t_z	
Conventional method	**19.25**	29.34	23.73	24.27	**13.06**	59.24	44.41
Proposed w/o absolute	21.56	27.48	23.30	**23.78**	13.89	47.92	34.68
Proposed	19.52	**26.38**	**22.04**	23.97	14.02	**45.87**	**27.71**

Table 2. Summary of experimental results using three types of scanning targets.

Method	AE (degree/mm)						Final drift (mm)
	θ_x	θ_y	θ_z	t_x	t_y	t_z	
Conventional method	126.46	260.33	130.52	137.21	**36.24**	210.46	117.30
Proposed w/o absolute	127.31	258.01	120.81	134.29	38.39	184.48	**90.78**
Proposed	**125.37**	**252.70**	**120.18**	**133.24**	40.63	**184.08**	96.46

where $\boldsymbol{p}_{i-1,i}$ and $\hat{\boldsymbol{p}}_{i-1,i}$ are the ground-truth and estimated relative pose between the frame $i-1$ and i, and M is the number of frames. We also evaluate the final drift, which is the distance between the ground-truth and estimated positions at the last frame. We compare the accuracy of the conventional method proposed by Prevost et al. [14]. Note that we implemented and trained this method according to [14] since there is no public implementation. We also evaluate the accuracy of the proposed method without absolute pose estimation to demonstrate the effectiveness of introducing absolute pose estimation.

4.2   Experimental Results

Table 1 shows the summary of experimental results using the breast phantom. The estimation accuracy is improved by RNN comparing the conventional method and the proposed method. The proposed method exhibits the best estimation accuracy in the methods as observed in the third and fourth row of Table 1 and the effectiveness of absolute pose estimation can be confirmed. Table 2 shows the summary of experimental results using three types of scanning targets. As in Table 1, the estimation accuracy of the proposed method is improved except for t_y. The errors for each parameter in Table 2 are larger than those in Table 1 since various movements of the probe include in the experiment

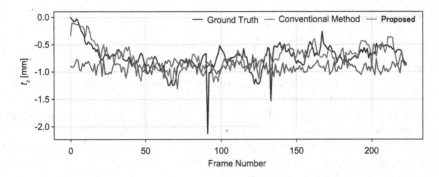

Fig. 3. Temporal variation of t_z estimated by each method.

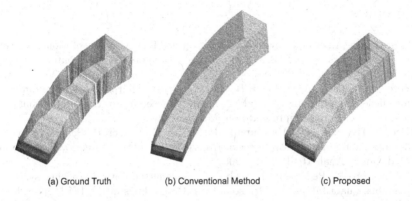

(a) Ground Truth   (b) Conventional Method   (c) Proposed

Fig. 4. Example of reconstructed US volume data: (a) ground truth, (b) conventional method, (c) proposed method.

of Table 2. Figure 3 shows the temporal variation of t_z estimated by each method. The conventional method cannot estimate large motion, and therefore shows the average temporal variation. The proposed method shows similar temporal variation to the ground truth, while it sometimes deviates significantly from it. Especially, the proposed method can track the ground-truth probe motion in the first half frame. Figure 4 shows the reconstructed US volume data using the probe position of the ground truth, the conventional method, and the proposed method. Each US volume data is reconstructed using StradView[3]. Since the conventional method estimates the average value of each parameter, the reconstructed US volume data is longer than the ground-truth US volume data. The proposed method exhibits better performance than the conventional method since the shape of the reconstructed volume data is similar to that of the ground truth.

---

[3] https://mi.eng.cam.ac.uk/Main/StradView.

5   Conclusion

In this paper, we proposed a 2D US probe pose estimation method only from US image sequences using deep learning. The proposed method employs the combination of CNN and RNN to estimate the relative and absolute probe pose. We also introduced the loss functions that take into account both relative and absolute pose to improve the estimation accuracy. Through a set of experiments using our dataset, we demonstrated that the proposed method exhibited better accuracy than the conventional method. In future work, we will apply the proposed method to US images of various parts and consider its application to sports medicine and point-of-care testing.

References

1. Dosovitskiy, A., et al.: FlowNet: learning optical flow with convolutional networks. In: Proceedings of the IEEE International Conference on Computer Vision, pp. 2758–2766, December 2015
2. Foret, P., Kleiner, A., Mobahi, H., Neyshabur, B.: Sharpness-aware minimization for efficiently improving generalization. In: Proceedings of the International Conference on Learning Representation, March 2021
3. Gee, A., Housden, R.J., Hassenpflug, P., Treece, G., Prager, R.: Sensorless freehand 3D ultrasound in real tissue: speckle decorrelation without fully developed speckle. Med. Image Anal. **10**, 137–149 (2006)
4. Goldsmith, A., Pedersen, P., Szabo, T.: An inertial-optical tracking system for portable, quantitative, 3D ultrasound. In: Proceedings of the IEEE International Ultrasonics Symposium, pp. 45–49, November 2008
5. Guo, H., Xu, S., Wood, B., Yan, P.: Sensorless freehand 3D ultrasound reconstruction via deep contextual learning. In: Martel, A.L., et al. (eds.) MICCAI 2020. LNCS, vol. 12263, pp. 463–472. Springer, Cham (2020). https://doi.org/10.1007/978-3-030-59716-0_44
6. Hastenteufel, M., Vetter, M., Meinzer, H.P., Wolf, I.: Effect of 3D ultrasound probes on the accuracy of electromagnetic tracking systems. Ultrasound Med. Biol. **32**(9), 1359–1368 (2006)
7. Horvath, S., et al.: Towards an ultrasound probe with vision: structured light to determine surface orientation. In: Linte, C.A., Moore, J.T., Chen, E.C.S., Holmes, D.R. (eds.) AE-CAI 2011. LNCS, vol. 7264, pp. 58–64. Springer, Heidelberg (2012). https://doi.org/10.1007/978-3-642-32630-1_6
8. Ito, S., Ito, K., Aoki, T., Ohmiya, J., Kondo, S.: Probe localization using structure from motion for 3D ultrasound image reconstruction. In: Proceedings of the International Symposium on Biomedical Imaging, pp. 68–71, April 2017
9. Krizhevsky, A., Sutskever, I., Hinton, G.E.: ImageNet classification with deep convolutional neural networks. In: Advances in Neural Information Processing Systems, vol. 25, pp. 1097–1105 (2012)
10. Lange, T., Kraft, S., Eulenstein, S., Lamecker, H., Schlag, P.: Automatic calibration of 3D ultrasound probes. In: Handels, H., Ehrhardt, J., Deserno, T., Meinzer, H.P., Tolxdorff, T. (eds.) Bildverarbeitung für die Medizin. INFORMAT, pp. 169–173. Springer, Heidelberg (2011). https://doi.org/10.1007/978-3-642-19335-4_36

11. Miura, K., Ito, K., Aoki, T., Ohmiya, J., Kondo, S.: Localizing 2D ultrasound probe from ultrasound image sequences using deep learning for volume reconstruction. In: Hu, Y., et al. (eds.) ASMUS/PIPPI -2020. LNCS, vol. 12437, pp. 97–105. Springer, Cham (2020). https://doi.org/10.1007/978-3-030-60334-2_10

12. Miura, K., Ito, K., Aoki, T., Ohmiya, J., Kondo, S.: Probe localization from ultrasound image sequences using deep learning for volume reconstruction. In: Proceedings of the International Forum on Medical Imaging in Asia, p. 103, January 2021

13. Prevost, R., et al.: 3D freehand ultrasound without external tracking using deep learning. Med. Image Anal. **48**, 187–202 (2018)

14. Prevost, R., Salehi, M., Sprung, J., Ladikos, A., Bauer, R., Wein, W.: Deep learning for sensorless 3D freehand ultrasound imaging. In: Descoteaux, M., Maier-Hein, L., Franz, A., Jannin, P., Collins, D.L., Duchesne, S. (eds.) MICCAI 2017. LNCS, vol. 10434, pp. 628–636. Springer, Cham (2017). https://doi.org/10.1007/978-3-319-66185-8_71

15. Rafii-Tari, H., Abolmaesumi, P., Rohling, R.: Panorama ultrasound for guiding epidural anesthesia: a feasibility study. In: Taylor, R.H., Yang, G.-Z. (eds.) IPCAI 2011. LNCS, vol. 6689, pp. 179–189. Springer, Heidelberg (2011). https://doi.org/10.1007/978-3-642-21504-9_17

16. Rousseau, F., Hellier, P., Barillot, C.: A fully automatic calibration procedure for freehand 3D ultrasound. In: Proceedings of the IEEE International Symposium on Biomedical Imaging, pp. 985–988, July 2002

17. Shi, X., Chen, Z., Wang, H., Yeung, D., Wong, W., Woo, W.: Convolutional LSTM network: a machine learning approach for precipitation nowcasting. In: Proceedings of the International Conference on Neural Information Processing Systems, pp. 802–810, December 2015

18. Stolka, P., Kang, H., Choti, M., Boctor, E.: Multi-DoF probe trajectory reconstruction with local sensors for 2D-to-3D ultrasound. In: Proceedings of the IEEE International Symposium on Biomedical Imaging, pp. 316–319, April 2010

19. Sun, S.-Y., Gilbertson, M., Anthony, B.W.: Probe localization for freehand 3D ultrasound by tracking skin features. In: Golland, P., Hata, N., Barillot, C., Hornegger, J., Howe, R. (eds.) MICCAI 2014. LNCS, vol. 8674, pp. 365–372. Springer, Cham (2014). https://doi.org/10.1007/978-3-319-10470-6_46

20. Xue, F., Wang, X., Li, S., Wang, Q., Wang, J., Zha, H.: Beyond tracking: selecting memory and refining poses for deep visual odometry. In: Proceedings of the International Conference on Computer Vision and Pattern Recognition, pp. 8575–8583, January 2019

Evaluation of Low-Cost Hardware Alternatives for 3D Freehand Ultrasound Reconstruction in Image-Guided Neurosurgery

Étienne Léger[1]([✉]), Houssem Eddine Gueziri[2], D. Louis Collins[2], Tiberiu Popa[1,3], and Marta Kersten-Oertel[1,3]

[1] Department of Computer Science and Software Engineering, Gina Cody School of Engineering and Computer Science, Concordia University, Montréal, QC, Canada
`etienne.leger@mail.concordia.ca`
[2] McConnell Brain Imaging Centre, Montreal Neurological Institute and Hospital, McGill University, Montréal, QC, Canada
[3] PERFORM Center, Concordia University, Montréal, QC, Canada

Abstract. The evolution of consumer-grade hardware components (*e.g.,* trackers, portable ultrasound probes) has opened the door for the development of low cost systems. We evaluated different low-cost tracking alternatives on the accuracy of 3D freehand ultrasound reconstruction in the context of image-guided neurosurgery. Specifically, we compared two low-cost tracking options, an Intel RealSense depth camera setup and the OptiTrack camera to a standard commercial infrared optical tracking system, the Atracsys FusionTrack 500. In addition to the tracking systems, we investigated the impact of ultrasound imaging on 3D reconstruction. We compared two ultrasound systems: a low-cost handheld ultrasound system and a high-resolution ultrasound mobile station. Ten acquisitions were made with each tracker and probe pair. Our results showed no statistically significant difference between the two probes and no difference between high and low-end optical trackers. The findings suggest that low cost hardware may offer a solution in the operating room or environments where commercial hardware systems are not available without compromising on the accuracy and usability of US image-guidance.

Keywords: Low cost · 3D freehand reconstruction · Ultrasound · Neurosurgery

1 Introduction

Image-guided neurosurgery (IGNS) systems have shown positive impacts on tailoring craniotomies, reducing interventional errors, increasing tumour resection percentages and improving patient survival rates. However, these systems suffer from accuracy degradation as a procedure progresses and the patient to image

© Springer Nature Switzerland AG 2021
J. A. Noble et al. (Eds.): ASMUS 2021, LNCS 12967, pp. 106–115, 2021.
https://doi.org/10.1007/978-3-030-87583-1_11

alignment computed at the beginning of surgery gets invalidated by the movement and deformation of the brain, or *brain shift*. To reregister the patient intraoperatively, updated images can be acquired, using either intraoperative magnetic resonance images (iMRI) (*e.g.,* Clatz *et al.* [4]), intraoperative computed tomography (iCT) (*e.g.,* Riva *et al.* [12]) or intraoperative ultrasound (iUS) (*e.g.,* Reinertsen *et al.* [11]). The latter is much less expensive, has a smaller footprint in the OR, and has shown usefulness in neurosurgery, for intraoperative image-based registration correction to account for brain shift [15].

Although more affordable than MRI and CT solutions, the price range of US systems still varies significantly from a low-cost handheld system ($\sim$2–8k USD) to a high-resolution station (between 50–250k USD). In addition to the intraoperative imaging modality, another hardware component used in IGNS to perform 3D freehand ultrasound reconstruction is the tracking system. This component, usually an optical tracking system, accounts for a substantial portion of the hardware costs of open-source IGNS systems. In this work, we compared the accuracy of ultrasound reconstruction obtained with different hardware setups at a broad range of price points. For the ultrasound transducer, two options were compared: a $\sim$250k USD mobile station and a $\sim$7k USD handheld system. For tracking four options are compared: a $\sim$25k USD high-end optical tracker, a $\sim$3k USD lower-end optical tracker, a sensor fusion hybrid tracking method which uses a $\sim$200 USD depth camera and a $\sim$20 USD equivalent RGB camera. Using these different setups we aimed to answer the following questions: Can compromises be made on some of the components without sacrificing too much on accuracy of the 3D freehand US reconstruction? If so which ones and how is a given budget best invested between these components?

2   Previous Work

Cenni *et al.* [3] looked at the effect of using different hardware setups on 3D freehand US reconstruction quality. They tested their method with two different optical tracking systems (exact models not disclosed) and report having found no noticeable difference in the reconstruction quality with the two systems. However, separate acquisitions were made independently with each tracker, making the comparison less robust.

The low cost alternative tracking method tested in our study is similar to that presented by Asselin *et al.* [1]. In their work, Asselin *et al.* developed a sensor fusion tracking method that uses a depth camera and an RGB camera to detect an ArUco marker in the RGB image to determine the x and y position in space and the depth camera to determine its z position. They found that the method worked very well, much better than when using an ArUco marker alone and would be suitable for intraoperative tool tracking. The present study thus extends their work in assessing if that low-cost hardware and method can be used for 3D freehand reconstruction. The final tracking method tested is that of using an ArUco marker alone. It serves as our baseline, similarly to Asselin *et al.*'s study.

A number of studies have investigated the reconstruction quality of different 3D freehand reconstruction algorithms. The interested reader is referred to Solberg *et al.* [14] and Rohling *et al.* [13] for a review in this area. A review of US probe calibration in the context of 3D freehand US reconstruction is also available in [9].

3   Methodology

We designed an experiment to simultaneously test the impact of hardware on the accuracy of 3D US reconstructed volumes, with both dimensional distortion as well as shape (angular) distortion in all three dimensions.

3.1   Hardware Setups

The acquisitions were made using two ultrasound probes in combination with four tracking systems. The tested ultrasound probes are: (1) the MicrUs MC4-2R20S-3 probe (TELEMED, Vilnius, Lithuania); and (2) the BK3500 14L3 probe (BK Medical, Peabody, MA, USA). The tracking systems that were tested are: (1) ArUco markers [5] captured with the RealSense RGB camera; (2) the RealSense D435 (Intel Corporation, Santa Clara, CA, USA); (3) the Optitrack V120:Duo (NaturalPoint Inc., Corvallis, OR, USA); and (4) the Atracsys Fusion-Track 500 (Atracsys LLC, Puidoux, Switzerland). All combinations of ultrasound and tracker were used to capture ultrasound acquisitions.

3.2   Phantoms and Marker Construction

To enable the most precise and fair assessment possible, a phantom and markers were designed for the experiment. A wire phantom built from Lego™ with eight wires pulled tautly through between Lego bricks was constructed (Fig. 1a). The wires form a cuboidal shape of precisely known dimensions. All wires cross perpendicularly, thus angles between line segments are precisely known. Lego bricks themselves are accurate to within 0.04 mm [7] and the wires were carefully pulled between them, which translates to a very accurate phantom. The cuboid measures 11.20 mm by 9.60 mm by 19.00 mm in x, y and z respectively. The phantom was immersed in water for US acquisition.

A custom marker, similar to that of Asselin *et al.* [1], was designed to enable all trackers (*i.e.*, both RGB camera and optical) to capture the position of the probe in the same coordinate frame (see Fig. 1b). The marker pivot (3D position) for all tracking methods was defined to be a common point at the center of the construction (the center of the ArUco marker, which corresponds to the centroid of the reflective sphere positions). The marker was 3D printed on a Raise 3D Pro2 printer (Raise 3D Technologies, Inc., Irvine, CA, USA) using a 0.1 mm layer height. A rigid probe attachment bracket was also designed and printed with the same printer settings. A similar marker was designed as a reference and attached to the phantom. This custom design and precise alignment of the tracked position for all trackers was done in order to reduce potential bias in the comparison.

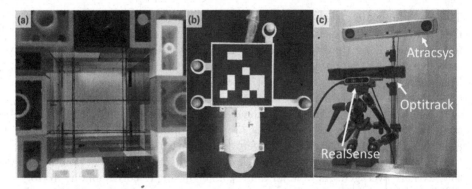

Fig. 1. Experimental setup: (a) Lego-wire phantom, (b) custom hybrid markers and probe attachment and (c) tracker arrangement for data acquisition.

3.3   Experiments

Tracking was captured simultaneously with all tracking systems for each US acquisition. Trackers were each placed at their optimal working distance from the scene to mimic a real-world scenario. Trackers were also each placed at as close as possible to the same viewing angle with respect to the scene, thus minimizing measurement volume as a confounding factor. The trackers were all aligned with the phantom so that the axes of the tracking volume would match that of the phantom (see Fig. 1c). To simplify the setup the live camera feed used for detecting the ArUco markers was that of the RealSense. This enabled us to have only three physical devices in the test setup while allowing testing with four tracking methods. The resolution of the RealSense RGB camera is 848 by 480 pixels, which is low for modern hardware. So, even though it was captured on a more expensive device, it could be achieve just as well with a $20 webcam.

The ultrasound probes were calibrated, both temporally and spatially, using fCal from PLUS (Public software Library for UltraSound imaging research) toolkit [6] (version 2.8). fCal implements the 3 N-wires calibration procedure [2], a method that was previously shown to be reliable and accurate [8]. This calibration was computed for each ultrasound probe with the tracker corresponding to the high-end of its price bracket (for the BK imaging system, the FusionTrack 500 tracker and for the Telemed imaging system, the V120:Duo tracker). The reasoning behind this was that using similarly priced devices in a system might be a more common use case.

Ten sweeps of the phantom were acquired with each ultrasound probe. For each acquisition, the tracking data was recorded simultaneously with all tracker systems. All sweeps were done in one linear motion done along the z axis. Independent reconstructions were then computed from each sweep and hardware combination, using the PLUS reconstruction [6]. Thus, from the 20 acquired sweeps, a total of 80 volume reconstructions were computed.

On each of these reconstructed volumes, the eight lines corresponding to all wires were manually segmented using 3D Slicer [10] version 4.11.20210226.

The order in which segmentation was performed for each trial was random-ized between conditions to reduce potential operator bias. All segmentations were performed by the same operator. The intersection of the eight line pairs corresponding to the corners of the reconstructed cuboid were computed in a least-square sense. These eight constructed points were then used in all further analysis. The distance between these points were used to compute the dimen-sions of the cuboid, or the dimensional distortion (DD) along each axis and angles between the line segments were used to compute angular distortion (AD) around each axis. All metrics were averaged over each axis. This means that for the dimensions, all four segments spawning from the connections of points along that axis are averaged. As well, for angles, all eight angles corresponding to rota-tion around that axis are averaged. Averaging is more robust and reduces the effect of uncertainty associated with segmentation. Finally, the total cuboid vol-ume was computed, which allows easier comparison with previous studies which computed the error as a percentage of volume.

4   Results

The image to probe calibration reprojection error for the Telemed system was 0.87 mm and for the BK system 1.26 mm. The temporal calibration yielded a 38 ms latency for the Telemed and a 48 ms one for the BK.

4.1   Reconstruction Quality Results

For both DD and AD, the absolute value of the error is used in analysis as both a negative or positive error would have similarly undesirable effects on the usability of the resulting reconstruction. Table 1 shows the DD results for all combinations of hardware. Table 2 shows the AD for all combination of hardware. DD is reported as a percentage error of the supposed length value and AD is reported as an angle difference from the supposed angle (90°). Results in both tables show the mean value for each setup with the standard deviation in parentheses.

We found that the probe used had little impact on the overall accuracy of the reconstruction. A two-way ANOVA revealed that the BK and Telemed reconstructions were not statistically significantly different from one another on neither dimensional nor angular error on almost any axis. They were only differ-ent in the x dimension, where the BK was worse than the Telemed ($p = 0.0275$), for all other metrics they were not statistically different. For that reason, data for both probes was bundled in Fig. 3. Reconstructions done with the Atracsys and Optitrack trackers were also not significantly different from one another on any metric and any dimension. All differences that were statistically different from the null hypothesis are labelled with stars in Fig. 3.

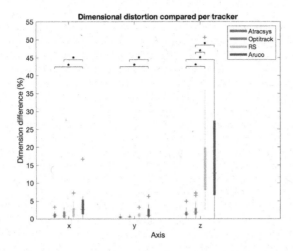

Fig. 2. Boxplots of dimensional distortions compared per tracker. Relationships marked with a star ($\star$) are those where the difference between group means are statistically significant to within $p < 0.05$. Devices are ordered in decreasing order of cost.

4.2   Qualitative Results

When visually inspecting the 3D ultrasound reconstructions a number of things can be seen. First, the Telemed and BK are clearly different in terms of image quality (Fig. 4). The wires appear more fuzzy in the images acquired with the Telemed. Second, there was very noticeable visual differences in reconstruction quality between volumes obtained with either the Atracsys or Optitrack and those obtained with ArUco or RealSense. In those from the ArUco alone or RealSense the wires are much less clearly defined (those of the ArUco alone being slightly worse). This lower visual quality of the reconstruction translated

Table 1. Mean dimensional distortion per axis for each combination of hardware as well as total volumetric error. Devices are ordered in decreasing order of cost.

Probe	Tracker	Dimensional distortion (%)			Volume (%)
		x-axis	y-axis	z-axis	
BK	Atracsys	1.37 ± 0.96	0.59 ± 0.33	1.77 ± 1.29	2.54 ± 2.16
	Optitrack	1.40 ± 1.05	0.66 ± 0.38	2.13 ± 2.16	3.30 ± 2.77
	RS	2.79 ± 2.32	0.82 ± 0.43	12.28 ± 6.53	12.75 ± 9.27
	ArUco	5.06 ± 4.80	1.90 ± 2.16	19.02 ± 10.63	20.73 ± 14.94
Telemed	Atracsys	0.78 ± 0.37	0.41 ± 0.39	1.28 ± 0.85	1.14 ± 0.89
	Optitrack	1.08 ± 1.09	0.47 ± 0.28	2.70 ± 1.80	3.44 ± 2.71
	RS	1.85 ± 1.22	1.36 ± 0.86	21.65 ± 13.72	22.53 ± 13.23
	ArUco	2.64 ± 1.96	1.50 ± 1.20	13.74 ± 13.87	13.83 ± 14.66

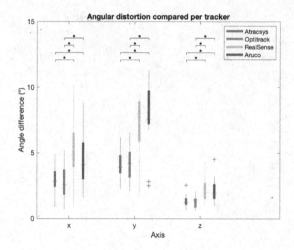

Fig. 3. Boxplots of angular distortions compared per tracker. Relationships marked with a star $(\star)$ are those where the difference between group means are statistically significant to within $p < 0.05$. Devices are ordered in decreasing order of cost.

very strongly when doing the manual segmentation. Wires in the ArUco and RealSense acquired volumes were more difficult to segment. They are much noisier and jagged, which made the segmentation process more error prone. Picking the center of those wires was harder on those reconstructions than those obtained with the other two systems.

5   Discussion

The fact that reconstructions made with the BK and the Telemed probe were visually different is not very surprising, image resolution and probe frequency of the BK are significantly higher, 728×892 and $12\,\mathrm{MHz}$ compared to 512×512 and $4\,\mathrm{MHz}$ for the Telemed. However, this difference did not translate into a measurable difference in reconstruction volume quality, meaning that the wires appeared more diffuse but their position corresponded. Even though images are noisier with a low cost probe, the reconstructed volume is still accurate, which leads us to believe it would perform reasonably well in brain-shift correction or for visualizing tool trajectories (*e.g.,* catheter, ventricular drain or needle) with appropriate user training. At the same time the jaggedness of the reconstructed edges on lower cost hardware might impact intraoperative registration given that the noise introduces artificial gradients. This will be explored in future work.

The fact that cheap hardware (both probe and optical tracker) works similarly to more expensive hardware hints that errors arising from other sources (*e.g.,* calibration, reconstruction, unevenness in the sweep acquisition movement) are higher than that of the measurements for all devices, even lower cost ones.

Table 2. Mean angular distortion per axis for each combination of hardware. Devices are ordered in decreasing order of cost.

Probe	Tracker	Angular distortion (°)		
		x-axis (pitch)	y-axis (yaw)	z-axis (roll)
BK	Atracsys	2.79 ± 0.95	3.31 ± 0.56	1.36 ± 0.57
	Optitrack	3.24 ± 1.16	3.36 ± 0.96	1.31 ± 0.48
	RS	5.24 ± 1.38	7.75 ± 2.44	2.29 ± 0.94
	ArUco	5.03 ± 2.12	9.65 ± 1.37	2.16 ± 0.65
Telemed	Atracsys	3.12 ± 1.17	4.82 ± 0.70	1.19 ± 0.35
	Optitrack	2.30 ± 1.31	5.00 ± 1.12	1.36 ± 0.44
	RS	5.00 ± 2.98	6.30 ± 2.38	2.13 ± 0.90
	ArUco	4.26 ± 1.87	6.53 ± 2.18	2.16 ± 1.09

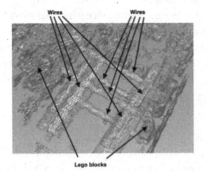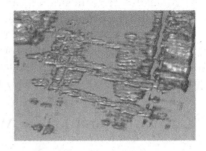

Fig. 4. Side-by-side comparison of typical reconstruction results obtained with each US acquisition systems. Both acquisitions depicted were acquired with the Atracsys tracking system. Left: BK imaging system; Right: Telemed imaging system.

In our experiment and in general in 3D freehand US reconstruction, the time difference between image timestamps and tracking timestamps is assumed to be fixed. The temporal calibration done prior to acquisition enables computing this time difference, which can then be compensated for in software upon data streams arrivals. However, it was observed that the BK latency fluctuated over time. For this reason, the BK was perhaps at a bit of a disadvantage. In our particular setup this fluctuation could have arisen from many sources: US system software, network card drivers, operating system or other receiving software. Latency should be considered with great care in this application and efforts should be made to ensure that the latency is not only as low as possible, but also, and very importantly, that it remains as constant as possible throughout an acquisition.

The difficulty described in the previous section in doing the manual segmentation on the cheapest tracking hardware has consequences beyond just the segmentation process itself. Not only is the process more time consuming for these acquisitions as the viewer takes longer to understand the US images but more importantly this leads to less accurate segmentation. This less accurate segmentation might be what causes both ArUco and RealSense to be indistinguishable statistically. Values are quite different for the z-axis, but the standard deviation on both samples is also large. There is a possibility that a genuine statistical difference between the two might be obfuscated by this segmentation difficulty due to low quality of the reconstructed volumes.

We found that all systems, even higher end ones, performed significantly worse in the z direction. This was expected, as all tracking methods tested, be it the commercial optical trackers or the experimental sensor-fusion method, are vision-based, meaning that they measure distances in images. They are therefore more accurate in the image plane than perpendicular to it. Although, and while all system suffer from this, the marker-based (ArUco and RealSense) were much more affected.

Finally, it is worth noting an important limitation in the design of our experiment. Manual wire segmentation, as performed in the experiment, allowed us to compensate for discontinuous data, especially when the quality of the reconstructed volume was low. Although this approach allows for capturing of the overall dimensional and angular distortions, local artifacts such as deformations and mis-reconstructions were attenuated. The effect of these artifacts on the outcome of an IGNS application need to be investigated.

6   Conclusion and Future Work

In this study, a phantom and protocol to measure 3D freehand US reconstruction distortion was presented. The wire and Lego phantom is easy and cheap to build and the protocol is easy to replicate. This allows for a more standardized comparison of reconstruction methods and tracking methods in the future. The protocol was used in a study to gain insight into the impact of different hardware components' cost on reconstruction accuracy. Four tracking systems were compared, whose cost were an order of magnitude apart from one another, as well as two US imaging systems that were roughly two orders of magnitude apart in price. We found that the cheapest US imaging system didn't yield reconstructions that were measurably worse than the high-end system. This thus suggests that for this application a cheap US imaging system may be used to reduce overall system cost. For tracking, the cheapest optical tracking system performed statistically the same as the high-end optical tracker. This shows the feasibility of using low cost hardware. However, the camera-based tracking methods performed significantly worse. To improve on the sensor fusion method, the depth camera could be used to track the shape of a marker in space, and this will be explored in future work. In future work as well, a second series of experiments will be performed to test more specifically how volume registration is impacted by the varying quality of reconstructions obtained with different hardware components.

References

1. Asselin, M., Lasso, A., Ungi, T., Fichtinger, G.: Towards webcam-based tracking for interventional navigation. In: Fei, B., Webster III, R.J. (eds.) Medical Imaging 2018: Image-Guided Procedures, Robotic Interventions, and Modeling, vol. 10576, pp. 534–543. International Society for Optics and Photonics, SPIE (2018). https://doi.org/10.1117/12.2293904
2. Carbajal, G., Lasso, A., Gómez, Á., Fichtinger, G.: Improving N-wire phantom-based freehand ultrasound calibration. Int. J. Comput. Assist. Radiol. Surg. **8**(6), 1063–1072 (2013). https://doi.org/10.1007/s11548-013-0904-9
3. Cenni, F., Monari, D., Desloovere, K., Aertbeliën, E., Schless, S.H., Bruyninckx, H.: The reliability and validity of a clinical 3D freehand ultrasound system. Comput. Methods Programs Biomed. **136**, 179–187 (2016). https://doi.org/10.1016/j.cmpb.2016.09.001
4. Clatz, O., et al.: Robust nonrigid registration to capture brain shift from intraoperative MRI. IEEE Trans. Med. Imaging **24**(11), 1417–1427 (2005). https://doi.org/10.1109/TMI.2005.856734
5. Garrido-Jurado, S., Muñoz-Salinas, R., Madrid-Cuevas, F.J., Marín-Jiménez, M.J.: Automatic generation and detection of highly reliable fiducial markers under occlusion. Pattern Recogn. **47**(6), 2280–2292 (2014). https://doi.org/10.1016/j.patcog.2014.01.005
6. Lasso, A., Heffter, T., Rankin, A., Pinter, C., Ungi, T., Fichtinger, G.: PLUS: open-source toolkit for ultrasound-guided intervention systems. IEEE Trans. Biomed. Eng. **61**(10), 2527–2537 (2014). https://doi.org/10.1109/TBME.2014.2322864
7. Lemes, S.: Comparison of similar injection moulded parts by a coordinate measuring machine. SN Appl. Sci. **1**(2), 1–8 (2019). https://doi.org/10.1007/s42452-019-0191-3
8. Mercier, L., et al.: New prototype neuronavigation system based on preoperative imaging and intraoperative freehand ultrasound: system description and validation. Int. J. Comput. Assist. Radiol. Surg. **6**(4), 507–522 (2011). https://doi.org/10.1007/s11548-010-0535-3
9. Mercier, L., Langø, T., Lindseth, F., Collins, D.L.: A review of calibration techniques for freehand 3-D ultrasound systems. Ultrasound Med. Biol. **31**(4), 449–471 (2005). https://doi.org/10.1016/j.ultrasmedbio.2004.11.015
10. Pieper, S., Halle, M., Kikinis, R.: 3D slicer. In: 2004 2nd IEEE International Symposium on Biomedical Imaging: Macro to Nano 1, pp. 632–635 (2004). https://doi.org/10.1109/isbi.2004.1398617
11. Reinertsen, I., Lindseth, F., Askeland, C., Iversen, D.H., Unsgård, G.: Intraoperative correction of brain-shift. Acta Neurochir. **156**(7), 1301–1310 (2014). https://doi.org/10.1007/s00701-014-2052-6
12. Riva, M., et al.: Intraoperative computed tomography and finite element modelling for multimodal image fusion in brain surgery. Operative Neurosurg. **18**(5), 531–541 (2019). https://doi.org/10.1093/ons/opz196
13. Rohling, R., Gee, A., Berman, L.: A comparison of freehand three-dimensional ultrasound reconstruction techniques. Med. Image Anal. **3**(4), 339–359 (1999). https://doi.org/10.1016/S1361-8415(99)80028-0
14. Solberg, O.V., Lindseth, F., Torp, H., Blake, R.E., Nagelhus Hernes, T.A.: Freehand 3D ultrasound reconstruction algorithms-a review. Ultrasound Med. Biol. **33**(7), 991–1009 (2007). https://doi.org/10.1016/j.ultrasmedbio.2007.02.015
15. Unsgaard, G., et al.: Intra-operative 3D ultrasound in neurosurgery. Acta Neurochir. **148**(3), 235–253 (2006). https://doi.org/10.1007/s00701-005-0688-y

Application Potential of Robot-Guided Ultrasound During CT-Guided Interventions

Josefine Schreiter[1]([✉]), Fabian Joeres[1], Christine March[2], Maciej Pech[2], and Christian Hansen[1]

[1] Otto von Guericke University Magdeburg, Magdeburg, Germany
josefine.schreiter@ovgu.de
[2] University Hospital of Magdeburg, Magdeburg, Germany

Abstract. CT-guided interventions are common practices in interventional radiology to treat oncological conditions. During these interventions, radiologists are exposed to radiation and faced with a non-ergonomic working environment. A robot-guided ultrasound (US) as a complementing imaging method for the purpose of needle guidance could help to overcome these challenges. A survey with 21 radiologists was made to analyze the application potential of US during CT-guided interventions with regard to anatomical regions to be scanned as locations of target lesions as well as specific situations during which US could complement CT imaging. The results indicate that the majority of respondents already applied US during CT-guided interventions for reasons of real-time imaging of the target lesion, organ, and needle movement as well as for lesions that are difficult to visualize in CT. Potential situations of US application were identified as out-of-plane needle insertion and puncturing lesions within the liver and subcutaneous lymph nodes. Interaction with a robot-guided US should be intuitive and include an improved sterility concept.

Keywords: CT-guided interventions · Ultrasound-guided interventions · Robot-guided ultrasound · Interventional radiology

1 Introduction

In the past, needle-based interventions under computed tomography (CT) imaging have become common practices in interventional radiology, among them drain placements, biopsies, and oncological therapeutic procedures like ablation or brachytherapy. The main purpose of CT imaging during these interventions is a reliable progress control, namely to safely guide the interventional instrument to the target structure without injuring nearby risk structures. General challenges arise during these CT-guided interventions, first and foremost, the radiation exposure for medical staff and patients [5,12]. Besides that, the complex work environment in the intervention room (IR) causes physiological challenges

© Springer Nature Switzerland AG 2021
J. A. Noble et al. (Eds.): ASMUS 2021, LNCS 12967, pp. 116–125, 2021.
https://doi.org/10.1007/978-3-030-87583-1_12

for the radiologist [2]. Due to the protective garments used to shield radiologists from radiation and the awkward body positions resulting thereof, interventional radiologists are faced with non-ergonomic working conditions which might lead to repetitive stress injuries [4]. Other than the two direct methods of radiation reduction - lowering the radiation output during a CT-guided interventions or improving radiation protection [5] - it might be beneficial to substitute or complement CT imaging by implementing a non-ionizing imaging method. Ultrasound (US) provides the advantages of good soft tissue contrast, image generation without causing radiation as well as it being a non-invasive procedure. Due to that, it could be a suitable imaging method. In addition, the ergonomic challenges could be reduced by automating individual steps of the workflow, for instance US screening of relevant body parts. Robotic US systems have been developed for several cases [10], most of them focusing on hand held US in combination with robotic needle insertion. However, little focus has been put on developing a US probe which is guided by a robotic arm during thoracic or abdominal CT-guided interventions.

This work intends to identify specific situations that could potentially benefit from robot-guided ultrasound which complements CT imaging during CT-guided interventions with regard to anatomical regions to be imaged. A complementing imaging method means that besides CT imaging, which is used during pre- and postprocedural workflow steps, US imaging is additionally applied for needle guidance. To answer this question, an online survey for radiologists was administered and analyzed.

Some research works have addressed the application of US to puncture various target lesions. Damm et al. [3] reported that in two third of cases in vivo, initial puncturing of abdominal malignancies was possible under US guidance and that half of the lesions showed a better visibility in US imaging compared to CT fluoroscopy. Moreover, Wu et al. [15] were able to show that procedure times of US-guided radiofrequency ablation for hepatocellular carcinomas were significantly shorter compared to those under CT-guidance without major difference in complete ablation and recurrence rate of the two groups. Robotic assistance in the form of a robot-guided US probe has been introduced by several research groups. Many focused their development on application scenarios in the field of vascular diagnosis such as the development of robotic systems that are able to autonomously measure diameters of abdominal aortic aneurysms [14] or screen for peripheral arterial diseases [6]. Other approaches focused on robotic US-guidance assistance in liver diagnosis [9] and abdominal scanning in purpose of radiation therapy [13]. Neither approach analyzed requirements of the use cases in a deeper manner, but rather, focused on technical implementations.

2   Methods

2.1   Survey Design

An online survey[1] was created using the open source survey tool LimeSurvey [8]. The survey consisted of five sections:

S1: Personal professional information
S2: Current clinical practice of US application during CT-guided interventions
S3: Potential use cases of US application during CT-guided interventions
S4: Human-robot interaction in the IR
S5: Concluding questions

The number of questions in S2 differed depending on whether respondents reported to have already used US during CT-guided interventions in current clinical practice, or not. In total, the survey consisted of 15 questions for respondents who reported to have already used US (of which five multiple choice, seven text fields, three rating scale matrices with free text fields for explanations) and 11 questions for respondents who reported to have no experience using US during CT-guided interventions (of which five multiple choice, four text fields, two rating scale matrices with free text fields for explanations). S1 contained questions about personal professional experiences. Questions within section S2 investigated reasons as to why respondents do not use US during CT-guided interventions or why they do use it. In addition, the radiologists who use US screening were questioned about the case anatomy, as well as challenges and suggestions they could offer in respect to handling the device. Questions in section S3 dealt with the potential of US application in regards to human anatomy and specific situations. Nine anatomical locations were included in the survey, located in the abdomen or thorax. These anatomical locations were selected to be relevant due to being potential locations of target lesions which are commonly punctured in CT-guided interventions. In S4, an example scenario of a CT-guided intervention was given including a robotic arm with an attached US probe which is able to independently approach the body surface. It was further explained, that US scanning of the surface is performed by controlling the robotic arm. Questions were asked about the applicability of different human-robot interaction (HRI) methods to control the robot (see Fig. 1). These included direct interaction (hand-guiding) (HRI-1), indirect interaction via joystick (telemanipulation) (HRI-2), and indirect interaction via hand gesture control (HRI-3). The survey was concluded with a free text field, to provide respondents with an opportunity to give general remarks, as well as to provide consent for further contact.

---

[1] https://limesurvey.ovgu.de/index.php/542932?lang=en.

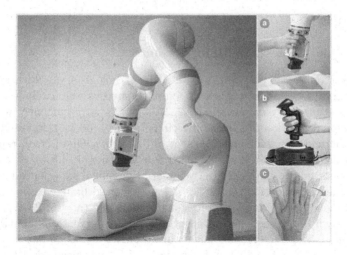

Fig. 1. HRI methods to control the robotic arm with attached US probe which were analyzed in the survey including hand-guiding (HRI-1) (a), telemanipulation via joystick (HRI-2) (b), and hand gesture control (HRI-3) (c).

2.2   Recruitment

The study data were obtained by distributing the aforementioned online survey to radiologists who are or have been practicing in the field of diagnostic and interventional radiology. This included radiologists from all departments of diagnostic and interventional radiology of university hospitals in Germany as well as former collaborators with this profession of our research group. The target group was further narrowed down by stating in the introductory text that persons who answer the survey should perform or should have performed needle-based CT-guided interventions on a regular basis. Participation in the online survey was anonymous and voluntary.

2.3   Data Analysis

Quantitative data were analyzed using the open source environment for statistical computing, R [11]. Qualitative data were analyzed by clustering answers of respondents into categories. Each category was ranked according to the total amount of answers within each category.

3 Results

3.1 Personal Professional Information

A total of 50 online surveys were answered by radiologists, of which 21 surveys were fully completed and analyzed. Incomplete surveys were excluded from the analysis. The majority of participating radiologists were currently working in the professional position of senior physicians (n = 14, 66.7%), followed by radiologists practicing as chief physicians (n = 5, 23.8%). One person reported to be a resident (n = 1, 4.8%), and another a fellow (n = 1, 4.8%). The majority of respondents said to have over ten years of professional experience in performing CT-guided interventions (n = 12, 57.1%), followed by five respondents who had five to ten years practical professional experience (n = 5, 23.8%) and two radiologists who had been performing CT-guided interventions for two to five years (n = 2, 9.5%). One respondent reported having less than one year of practical experience (n = 1, 4.8%) and one answer was not specified. Two thirds of the respondents answered to have performed over 500 CT-guided interventions throughout their professional career (n = 14, 66.7%), whereas four respondents completed 200–500 (n = 4, 19%), and three radiologists less than 100 CT-guided interventions (n = 3, 14.3%).

3.2 Current Clinical Practice

More than half of the respondents indicated to have experience applying US during CT-guided interventions as complementing imaging (n = 12, 57.1%), whereas the remaining radiologists (except one unspecified answer) had not used US for this purpose (n = 8, 38.1%).

Reasons for Non-usage. Four out of eight radiologists stated that CT imaging adequately fulfilled the purpose of visualizing the target lesion and the surrounding anatomy, and therefore never felt the need for a complementing or alternative imaging method. One fourth of this respondent group explained that they do not see advantages in the combined application of CT and US, and do not expect to save time or reduce radiation exposure by using this method. Furthermore, one radiologist supported his/her decision with expected inferior image quality and another radiologist with sterility issues of the US device.

Reasons for Usage. Seven out of 12 respondents stated that they have been applying US during CT-guided interventions for reasons of real-time imaging of relevant objects like the target lesion and/or the interventional instrument, followed by six radiologists who reported to be using US due to the superior imaging of the target structure, in case of lesions difficult to image in CT, or for the identification of risk structures. Further responses for usage were the reduction in radiation exposure, out-of-plane needle insertion, and the benefit of overall time reduction.

Quantitative results regarding US application for certain anatomical structures during CT-guided interventions in current clinical practice can be seen in Fig. 2a. Some of the respondents gave explanations for their ratings. These were clustered into categories (see Table 1).

Respondents reported about general challenges of applying additional US during CT-guided interventions. Eight respondents stated that the US device complicates the IR setting because of the additional equipment needed, higher space demand, and sterility issues. Additional challenges reported by the radiologists include handling the US device, and an increase in the time spent during procedures. Suggestions for an improved handling include a better integration of the US device in the IR (e.g. attachment of probe to CT housing, wireless probe, multifunctional display), intuitive probe handling, automatic scanning of the target region, as well as live image fusion of CT and US images.

3.3   Potential Use Cases

Respondents ranked the application potential of anatomical structures on a rating scale matrix (see Fig. 2b). Explanations about rankings were given by some of the radiologists which were clustered for analysis (see Table 1).

Radiologists reported that US imaging could be useful for puncturing superficial lesions, in case of out-of-plane needle insertion, targeting small, hard, and rounder anatomical structures, targeting moving lesions and identifying risk structures. Six respondents highlighted liver puncture with a high potential for US-guidance, with two respondents specifically stating liver lesions located in the hilus and with subphrenic position.

3.4   HRI in the IR

Respondents were asked to rate different interaction methods with a robot-guided US probe (6-DOF robotic arm) after giving them a theoretical scenario of a CT-guided intervention (see Fig. 2c). Supportive explanations of positive ratings (*realistic* and *very realistic*) of HRI-1 and HRI-2 were the expected intuitive and direct control of HRI-1 and the assumed intuitiveness of HRI-2. However, answers indicated concern in terms of sterility, expected inaccurate control, and cumbersome handling of HRI-1 as well as expected error-prone control of HRI-2. HRI-3 was rated with the least expected applicability with five respondents assuming the control to be inaccurate and one expecting the method to be *unrealistic* due to lack of feedback. Two radiologists stated that they could not give explanation for their rating due to their limited practical experience with the interaction methods.

Table 1. Answers of free text fields of the survey regarding the application potential of US during CT-guided interventions which were clustered into categories. The numbers indicate the total amount of answers given in the survey of section S2 and section S3.

	Lung	Liver	Kidney	Pancreas	Gall Bladder	Spleen	Lymph Nodes	Bones	Interstitium
High application potential of US due to/in case of ...									
Good anatomical visibility	–	1	–	–	4	–	–	–	–
Inferior CT visibility	–	2	2	–	1	–	2	–	1
Superior real-time imaging	–	3	3	–	1	1	3	–	–
Angulated trajectory	–	3	2	–	1	1	2	–	–
Superior visibility of risk structures	–	1	1	–	1	1	1	–	1
Low application potential of US due to...									
Poor anatomical visibility	3	–	–	1	2	1	–	2	–
Sufficient information generated with	6	1	1	5	3	1	2	4	1
CT imaging	–	1	2	1	1	4	5	1	5
Generally little punctured anatomy	–	–	–	5	1	3	–	3	–
Specific potential use cases	pleural lesions (2), pleural (1), peripher lesions (1)	–	–	–	–	–	subcutaneous (2), subcutaneous (1), axillary (1)	imaging trajectory (1), imaging trajectory (1)	subcutaneous (1), mobile lesions (1)

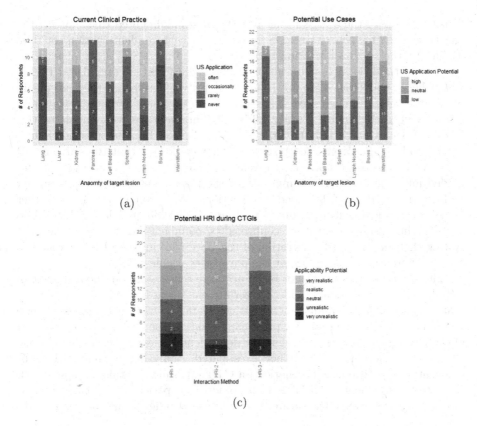

Fig. 2. Survey results regarding the application of US during CT-guided interventions in current clinical practice (a), potential use cases (b) and potential HRI with a robot-guided US probe (c).

4   Discussion

The majority of radiologists reported to have been using US alongside CT imaging during CT-guided interventions with the main motivation being the advantage of real-time imaging of moving target lesions (e.g. kidney due to breathing, lymph nodes) as well as superior imaging quality of certain anatomical structures difficult to outline under CT imaging. As both modalities share the characteristic of real-time imaging, we assume that US is preferably used in cases of poor soft tissue contrast in intraprocedural CT images. This assumption can be confirmed by the fact, that the soft tissue contrast in US is superior, especially for identification of lesions within abdominal organs [1]. This seems to be especially the case for target lesions which are located within the liver, kidney and lymph nodes as these anatomical regions were rated with the highest potential to be scanned with US during CT-guided interventions. Although some anatomies show bad visibility in US images for reasons such as artifacts caused by air, high tissue density, or due to the organs location, radiologists highlighted that imaging the anatomy of the trajectory with US for e.g. risk structure identification could be useful. Another main motivation for radiologists to use US as a complementing imaging method, is in case of out-of-plane needle insertion (angulated trajectory). Those trajectories are not in an axial body plane, hence they are not in plane with the CT image slice. In these cases, detection of the needle tip under CT imaging is challenging [7]. For CT-guided interventions with angulated trajectory, US is a potentially superior imaging method because it is not bound to any body plane. Even though, research has confirmed that the outcomes of US-guided interventions in some cases are comparable or better to those of CT-guided interventions [3,15], a minority of the respondents had not yet applied additional US during CT-guided interventions. We assume that this is rather an issue of clinical routine, training, and poor integration of the US device into the IR. This assumption can be confirmed by comments made by respondents, stating that an improved integration of the US device into the IR could lead to a higher potential of using US for the purpose of needle guidance.

The survey results might not be an accurate representation due to the small sample size of respondents, and the recruitment being limited to a single geographical location. In addition, the strict allocation of target lesions to nine anatomical regions does not give precise information about application potential of US for individual cases. Therefore the dependency of lesion location within one anatomical region has to be further investigated. Analyzing possible HRI methods contained a theoretical scenario of a CT-guided intervention. It is assumed, that most respondents have little to no practical experience with the proposed interaction methods. In the future, practical user studies should be conducted.

5   Conclusion

In this paper, we have investigated the application potential of robot-guided US for CT-guided interventions. The findings of the survey with radiologists indicate that US imaging has the highest potential of being applied for puncturing lesions within the liver and subcutaneous lymph nodes. Furthermore, there is a high application potential for out-of-plane needle insertions, and anatomical regions which have inferior soft tissue contrast in intraprocedural CT images. Radiologists emphasized the characteristic of US to enable real-time imaging of the target lesion, organ, and needle movement without radiation exposure to be highly beneficial. Criteria for a potential application of a robot-guided US regarding the integration into the IR, are an intuitive interaction method, a space saving integration and an improved sterility concept.

Acknowledgments. This work was funded by the Federal Ministry of Education and Research within the Forschungscampus *STIMULATE* under grant number 13GW0473A.

References

1. Braak, S.J., van Strijen, M.J.L., van Leersum, M., van Es, H.W., van Heesewijk, J.P.M.: Real-time 3D fluoroscopy guidance during needle interventions: technique, accuracy, and feasibility. AJR Am. J. Roentgenol. **194**(5), W445–W451 (2010)
2. Cornelis, F.H., et al.: Ergonomics in interventional radiology: awareness is mandatory. Medicina **57**(5), 500 (2021)
3. Damm, R., et al.: Ultrasound-assisted catheter placement in CT-guided HDR brachytherapy for the local ablation of abdominal malignancies: initial experience. In: RoFo: Fortschritte auf dem Gebiete der Rontgenstrahlen und der Nuklearmedizin, vol. 191, no. 1, pp. 48–53 (2019)
4. Dixon, R.G., et al.: Society of interventional radiology: occupational back and neck pain and the interventional radiologist. J. Vasc. Intervent. Radiol. (JVIR) **28**(2), 195–199 (2017)
5. Elsholtz, F.H.J., et al.: Radiation exposure of radiologists during different types of CT-guided interventions: an evaluation using dosimeters placed above and under lead protection. Acta Radiol. **61**(1), 110–116 (2020). Stockholm, Sweden (1987)
6. Kaschwich, M., von Haxthausen, F., Aust, T., Ernst, F., Kleemann, M.: Roboterbasierte ultraschallsteuerung. Gefässchirurgie **25**(5), 345–351 (2020). https://doi.org/10.1007/s00772-020-00670-z
7. Komaki, T., et al.: Robotic CT-guided out-of-plane needle insertion: comparison of angle accuracy with manual insertion in phantom and measurement of distance accuracy in animals. Eur. Radiol. **30**(3), 1342–1349 (2020). https://doi.org/10.1007/s00330-019-06477-1
8. LimeSurvey Project Team/Carsten Schmitz: Limesurvey: an open source survey tool (2012). http://www.limesurvey.org
9. Mustafa, A.S.B., et al.: Development of robotic system for autonomous liver screening using ultrasound scanning device. In: IEEE International Conference on Robotics and Biomimetics (ROBIO), Piscataway, NJ, pp. 804–809. IEEE (2013)
10. Priester, A.M., Natarajan, S., Culjat, M.O.: Robotic ultrasound systems in medicine. IEEE Trans. Ultrason. Ferroelectr. Freq. Control **60**(3), 507–523 (2013)
11. R Core Team: R: a language and environment for statistical (2021). https://www.R-project.org
12. Rathmann, N., et al.: Evaluation of radiation exposure of medical staff during CT-guided interventions. J. Am. Coll. Radiol. (JACR) **12**(1), 82–89 (2015)
13. Seitz, P.K., Baumann, B., Johnen, W., Lissek, C., Seidel, J., Bendl, R.: Development of a robot-assisted ultrasound-guided radiation therapy (USgRT). Int. J. Comput. Assist. Radiol. Surg. **15**(3), 491–501 (2020). https://doi.org/10.1007/s11548-019-02104-y
14. Virga, S., et al.: Automatic force-compliant robotic ultrasound screening of abdominal aortic aneurysms. In: 2016 IEEE/RSJ International Conference on Intelligent Robots and Systems (IROS), 9–14 October 2016, pp. 508–513. IEEE (2016)
15. Wu, J., Chen, P., Xie, Y.G., Gong, N.M., Sun, L.L., Sun, C.F.: Comparison of the effectiveness and safety of ultrasound-and CT-guided percutaneous radiofrequency ablation of non-operation hepatocellular carcinoma. Pathol. Oncol. Res. POR **21**(3), 637–642 (2015). https://doi.org/10.1007/s12253-014-9868-5

Classification and Image Synthesis

Towards Scale and Position Invariant Task Classification Using Normalised Visual Scanpaths in Clinical Fetal Ultrasound

Clare Teng[1]([✉]), Harshita Sharma[1], Lior Drukker[2], Aris T. Papageorghiou[2], and J. Alison Noble[1]

[1] Institute of Biomedical Engineering, University of Oxford, Oxford, UK
`clare.teng@eng.ox.ac.uk`
[2] Nuffield Department of Women's and Reproductive Health, University of Oxford, Oxford, UK

Abstract. We present a method for classifying tasks in fetal ultrasound scans using the eye-tracking data of sonographers. The visual attention of a sonographer captured by eye-tracking data over time is defined by a scanpath. In routine fetal ultrasound, the captured standard imaging planes are visually inconsistent due to fetal position, movements, and sonographer scanning experience. To address this challenge, we propose a scale and position invariant task classification method using normalised visual scanpaths. We describe a normalisation method that uses bounding boxes to provide the gaze with a reference to the position and scale of the imaging plane and use the normalised scanpath sequences to train machine learning models for discriminating between ultrasound tasks. We compare the proposed method to existing work considering raw eye-tracking data. The best performing model achieves the F1-score of 84% and outperforms existing models.

Keywords: Eye-tracking · Fetal ultrasound · Time-series classification · Visual scanpath

1 Introduction

During routine fetal ultrasound scans, sonographers are required to capture and store standard imaging planes of fetal anatomy [15]. These imaging planes are referred to as *anatomy planes*. Each anatomy plane is considered a separate *task*, for example brain and heart. To distinguish between the tasks, we use eye-tracking data of the sonographers recorded while they performed the scan. The eye-tracking data contains gaze information that allows us to analyse the sonographer's visual attention and scanpath during different parts of the scan, where a scanpath is the path taken by the observer when observing a scene.

Using eye-tracking data for fetal ultrasound task classification is challenging for several reasons. The dynamic movement of a fetus means that there are numerous ways to find and capture an anatomy plane. As sonographers gain more

© Springer Nature Switzerland AG 2021
J. A. Noble et al. (Eds.): ASMUS 2021, LNCS 12967, pp. 129–138, 2021.
https://doi.org/10.1007/978-3-030-87583-1_13

experience, they typically capture anatomy planes more quickly and efficiently, resulting in fast transitions between planes. The size of the anatomy on the screen is also dependent on what scale the sonographer views the image at. Due to these changes in scale and position (Fig. 1), the scanpaths associated with different tasks are not easily separable using simple discriminatory methods. Knowing that anatomy planes have unique anatomical landmarks [5], we are motivated to understand whether we can distinguish between the visual scanpaths of different scanning tasks. When considering skill assessment of full-length scans, being able to classify different scanning tasks at a given time is important. The aim of our work is to understand if eye-tracking data is sufficient for the identification of the fetal ultrasound task being performed.

Related Work. Current works using scanpaths to classify tasks use static representations of eye-tracking data, for example number of fixations and fixation duration [9]. Other works either analyse tasks which use a single image such as reading [7] or generate a static representation by superimposing the overall task-specific scanpath onto an image. Studies using scanpaths consider only a handful of entry points to reach their target [1] or use saccadic information for classification [8]. Such works are less suitable for our application because of the numerous ways a sonographer can capture an anatomy plane image (Fig. 1), and the uncertainty in identifying saccadic movement accurately. Other studies [2,4,18] which use scanpaths in videos utilise images as a data source. However, it is expensive to train models on image data, and our work only considers eye-tracking which is more computationally efficient.

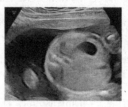 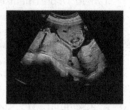

(a) Abdomen positioned on the right.

(b) Abdomen positioned in the top right hand corner.

Fig. 1. Example of difference in position and scale of an abdomen plane scan, where the image in a and b differ in both scale and position.

Contribution. Our main contributions are the following. We propose a feature engineering method using eye-tracking data that is able to account for the change in scale and position of anatomy during the scan. We compare different time-series classification models and use the best-performing model based on Gated Recurrent Units (GRU) to perform task classification using visual scanpaths for fetal ultrasound tasks.

2 Method

We present our proposed method for normalising scanpath with respect to scale and position, and our proposed time-series classification model for differentiating scanpaths of sonographers when searching for different anatomical structures.

Scanpath Normalisation for Scale and Position Invariance. Raw gaze points recorded by the eye-tracker along the x and y axis with respect to the screen dimensions of 1980 × 1080 pixels are defined as G_x, G_y. Raw gaze points normalised by screen dimensions of 1980 × 1080 pixels are defined as G_{xs}, G_{ys}, calculated as $G_{xs} = \frac{G_x}{1980}$, $G_{ys} = \frac{G_y}{1080}$.

To provide the model with information of gaze point position relative to the image, we normalise the gaze points with respect to the circumference of the anatomy. We manually draw bounding boxes using [14] around the circumference of the anatomy plane on a *cropped* (1008 × 784 pixels) image (shown as the red box Fig. 2). We exclude all text and clipboard images to view the circumference clearly. Then, we normalise the gaze points with respect to the corner positions of the bounding box along the x and y axis: X_L, X_R, Y_T, Y_B where L, R, T and B represent left, right, top and bottom, and the X and Y offsets (shown as X_{offset}, Y_{offset} in Fig. 2) created by using the cropped image, 427 and 66 pixels, respectively. An example of this normalisation process is shown in Fig. 2. Raw gaze points normalised by co-ordinates of a hand drawn bounding box on the image are given as G_{xBB}, G_{yBB} (Eq. 1). An example of drawn bounding boxes for the abdomen, brain and heart anatomical structures using the cropped image is shown in Fig. 3.

$$G_{xBB} = \frac{G_x - X_L - X_{offset}}{X_R - X_L} \quad \text{and} \quad G_{yBB} = \frac{G_y - Y_B - Y_{offset}}{Y_T - Y_B} \tag{1}$$

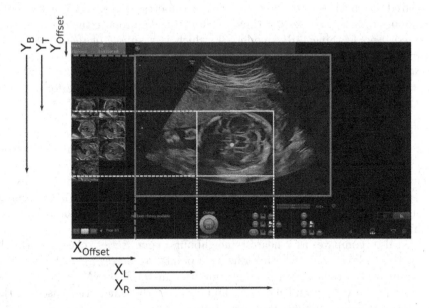

Fig. 2. An example showing how a raw gaze point (green) with co-ordinates G_x, G_y is normalised with respect to the hand drawn bounding box (yellow). The point of origin of the bounding box is given as the bottom left corner. (Color figure online)

We use the bounding box to capture the difference in scale when viewing the image by calculating the ratio of the screen that the anatomy occupies; a ratio of 1 is where the anatomy image occupies the entire screen. The area of screen occupied by the bounding box (yellow box, Fig. 2) divided by the area of the cropped image (red box, Fig. 2) is given as A. Our generated features are: G_{xBB}, G_{yBB}, A.

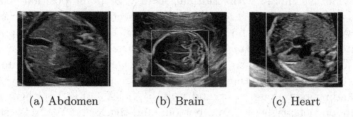

(a) Abdomen   (b) Brain   (c) Heart

Fig. 3. Example of manually drawn bounding boxes (in yellow) using [14] for cropped abdomen, brain and heart plane images. (Color figure online)

2.1 Time-Series Classification of Scanpaths

Time-series classification can be generalised into several categories. We focus on generative model-based methods as a baseline based on [1], which considers the joint distribution of data such as a hidden Markov model (HMM). We also consider whole series comparisons as another baseline, where each time-series is compared to another using a chosen distance metric, for example a k-nearest neighbours time series classifier (k-NN TSC) [10]. We propose the use of a standardised deep learning GRU model [3], which is a subset of recurrent neural networks (RNN) that retains time dependencies between sequences. Long short-term memory (LSTM) models are also a subset of RNNs and are similar to GRUs. However, GRUs have been shown to return comparable performance to the LSTM while requiring less specified parameters [20].

Baseline Comparisons. We compare our method to [1], as they classify video clips of surgical tasks using eye-tracking data. They use a k-means clustering algorithm to convert raw eye-tracking data into a discrete sequence consisting of cluster membership numbers. For each task, a HMM was trained, and each test sequence is scored against task specific HMMs, where the predicted class is selected as the model which returns the highest logarithmic likelihood. We refer to this model as *HMM*.

We also compare a k-nearest neighbours time series classifier (k-NN TSC) [10], to investigate whether raw gaze points are better for task classification compared to the coarse representation used in *HMM*. *k-NN TSC* calculates the distance between each time series and classifies the time-series based on the class most common amongst its k neighbours.

Parameter Selection. For *HMM*, we use the elbow method [19] to determine the optimal number of clusters for k-means, and a Gaussian HMM with a full covariance matrix. k = 5. For k-NN TSC, we use dynamic time warping as our distance metric based on previous works [2]. k = 2. For our GRU model, we use RayTune's [12] asynchronous successive halving algorithm (ASHA) [11] to tune our hyperparameters. The hyperparameters are: 32 hidden layers, 2 recurrent layers, 250 epochs, batch size = 4, dropout = 0.55 using an Adam optimiser with a learning rate of 0.003, and a cross entropy loss function.

3 Data

The dataset used are second-trimester manually labelled abdomen, brain, and heart planes as described in detail in [6]. The labelled data set contains the last 100 frames just before the sonographer freezes the video to take measurements of the captured anatomy plane [6]. A clip is defined as a single instance where the anatomy plane was searched for during a scan. We use sonographer eye-tracking gaze data (Tobii Eye Tracker 4C) sampled 90 Hz.

We choose the abdomen, brain, and heart plane scans because the sonographers spent the majority of their time viewing these planes [16,17], and abdomen and brain planes are considered easier to search for compared to the heart due to differences in anatomy size. Hence, we would expect the scanning characteristics between these anatomies to be distinct from each other. In total, there are 84, 160, 122 abdomen, brain and heart plane clips respectively. Our dataset consisted of 10 fully qualified sonographers carrying out the ultrasound scan, and 76 unique pregnant women as participants.

Clips which were shorter than 100 frames because the difference in time between the previous frozen segment and the next was less than 100 frames were zero padded creating equal length time-series of 100 gaze points. Any missing gaze points due to sampling discrepancies or tracking errors were interpolated.

4 Results

To increase the size of our dataset, we augmented the images by flipping the images about the horizontal, vertical, and horizontal and vertical axis. We also randomly down sampled the data with respect to the minority class, to prevent bias towards the majority classes. Our final dataset has 336 abdomen, brain and heart plane clips each. For training and testing we performed a 3-fold stratified cross validation. We used 80% for cross validation and 20% for tuning model parameters.

We used different sets of features (shown in Table 1) to demonstrate our proposed method of using bounding boxes for normalisation and proposed model performs better than current baseline models. Our classification results are shown in Table 1. For ease of reference, we used an affix 'Affix' column of Table 1 to refer to the corresponding features used; raw for raw gaze points, scr for gaze points normalised by screen dimensions, bb for gaze points normalised by bounding box.

Table 1 shows that our proposed feature engineering method to normalise raw eye-tracking data and model performs better than previous works [1] and several baselines, returning a weighted F1 score of 0.84.

Table 1. Comparison of weighted F1 scores and accuracies calculated using *HMM* [1], *k-NN TSC*, GRU to classify abdomen, heart and brain plane clips. Affix refers to the abbreviation for features used.

Model	Affix	Features	Weighted-F1	Accuracy
HMM	raw	G_x, G_y	0.38 ± 0.20	0.49 ± 0.19
	scr	G_{xs}, G_{ys}	0.38 ± 0.07	0.45 ± 0.09
k-NN TSC	raw	G_x, G_y	0.57 ± 0.05	0.57 ± 0.06
	scr	G_{xs}, G_{ys}	0.55 ± 0.04	0.55 ± 0.04
	scr+A	G_{xs}, G_{ys}, A	0.52 ± 0.05	0.54 ± 0.04
	bb+A	G_{xBB}, G_{yBB}, A	0.63 ± 0.03	0.64 ± 0.02
GRU	raw	G_x, G_y	0.56 ± 0.05	0.57 ± 0.05
	scr	G_{xs}, G_{ys}	0.68 ± 0.04	0.67 ± 0.04
	scr+A	G_{xs}, G_{ys}, A	0.72 ± 0.05	0.72 ± 0.05
	bb+A	G_{xBB}, G_{yBB}, A (ours)	$\mathbf{0.84 \pm 0.01}$	$\mathbf{0.83 \pm 0.01}$

Table 1 shows [1] is unable to classify fetal ultrasound tasks, where HMM(raw) and HMM(scr) returns score metrics between 38% and 49%. Instead, using k-NN TSC and GRU models improves the task classifier's performance by at least 20% - HMM(raw) and HMM(scr) versus k-NN TSC(raw) and k-NN TSC(scr), GRU(raw) and GRU(scr) respectively.

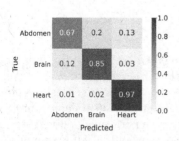

Fig. 4. Confusion matrix of our GRU model normalised with respect to total number of clips per anatomy plane in the test set (106 clips).

Normalisation using the bounding box shows an improvement of at least 10% (Table 1 GRU(scr+A) versus GRU(bb+A), k-NN TSC(scr+A) versus k-NN TSC(bb+A)), returning a final F1 score of 84%. There is a slight decrease (1%–3%) in performance when including the size of the anatomy relative to the screen for models k-NN TSC, but a slight increase (4%) for GRU using scr and scr+A. GRU is better able to use the anatomy size information compared to k-NN TSC. Overall, normalising gaze points with respect to the anatomy circumference is more relevant of task type, compared to how much space the anatomy occupies on the screen.

The confusion matrix for our GRU(bb+A) model is shown in Fig. 4. Figure 4 shows that abdomen scanpaths are more likely to be confused with brain and heart scanpaths, where 13% and 20% of abdomen scanpaths are misclassified as heart and brain scanpaths. Brain scanpaths are also more likely to be confused with abdomen scanpaths, where 12% are predicted as abdomen scanpaths. Heart scanpaths are most distinct, where only 3% are misclassified.

Class Imbalance Models. Since our initial dataset was unbalanced, where we had the most number of clips available for brain scanpaths and the least for abdomen scanpaths, we ran the GRU(bb+A) model using the focal loss [13] function which accounts for class imbalance when training the model. We compared the results using our original cross entropy loss which did not. We also compared the effect of augmenting our dataset.

Table 2. Weighted F1 scores and accuracies using our proposed GRU model comparing the original, downsampled (DS) and augmented (Aug) datasets. (i) original, (ii) original downsampled (iii) original, augmented and downsampled (our proposed model in Table 1) and (iv) original, augmented datasets for classification of abdomen, brain and heart planes.

Dataset	Loss function	Weighted-F1	Accuracy
(i) Original	Focal loss	0.81 ± 0.01	0.81 ± 0.02
(ii) Original + DS	Cross entropy	0.79 ± 0.04	0.78 ± 0.05
(iii) Original + Aug + DS (proposed)	Cross entropy	**0.84 ± 0.01**	**0.83 ± 0.01**
(iv) Original + Aug	Focal loss	0.83 ± 0.01	0.81 ± 0.02

Our results in Table 2 show that using an augmented dataset does not affect the performance when comparing the balanced (iii) and imbalanced (iv) datasets. The effect of using the downsampled dataset is seen when considering a smaller dataset (ii) (drop of 2–3%). Using focal loss returns more consistent results than that of using cross entropy, where the original dataset (i) returns a lower standard deviation across folds compared to the downsampled dataset (ii). Overall, for our application, using an augmented downsampled dataset did not affect the performance of our model negatively (iii) and (iv), but increasing the size of our dataset through augmentation improved performance by 4–5%.

5   Discussion

A qualitative investigation was performed to understand why brain and heart scanpaths are more likely to be confused with abdomen scanpaths, and why abdomen and brain scanpaths are more misclassified with each other compared to the heart. We show the qualitative investigation plot for abdomen scanpaths (Fig. 5) since they show the highest percentage of misclassification. Figure 5 shows a contour density plot (left) of training and correctly classified abdomen plane gaze points G_{xBB}, G_{xBB} with cumulative density masses at 4 equally spaced levels 0.2, 0.4, 0.6, 0.8 where 0.2 is the outer most contour and 0.8 is the inner most contour. The bi-variate distribution was calculated by superimposing a Gaussian kernel on each gaze point and returning a normalised cumulative sum. Figure 5 also shows abdomen scanpaths (right) which were incorrectly classified as a brain (orange) or heart (blue) scanpath.

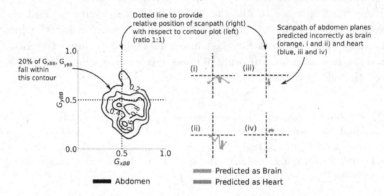

Fig. 5. Contour density plot of G_{xBB}, G_{xBB} (left), and abdomen scanpaths which were incorrectly classified as brain (orange, i and ii) or heart (blue, iii and iv) scanpath represented (right). (Color figure online)

Figure 5 shows that for abdomen planes, sonographer scanpaths are concentrated within the central area of the anatomy. However, for abdomen scanpaths predicted as heart, the sonographer focused on a single area (Fig. 5, iii and iv) similar to how sonographers visually search for the heart. For scanpaths predicted as brain, the sonographer moved the probe, causing their gaze to shift accordingly with the image (ii), or had moved their gaze across the screen (i) similar to how sonographers search for the brain.

For misclassified brain scanpaths, the image was small and occupied <50% of the screen, and the sonographer did not focus along the midline horizontally but diagonally across the plane. Misclassified heart scanpaths showed that the image itself was moving, indicating that the probe was moving, causing the sonographer to shift their gaze accordingly or the sonographer was looking around the walls of the heart cavity.

6   Conclusion

In this paper, we have presented a method for normalising eye-tracking data with respect to the circumference of the anatomy which is able to account for changes in position and scale of the anatomy image during the scan. With our method, we have improved task classification using eye-tracking data score metrics by at least 15% compared to other methods. We also present a GRU model which performed better than other classification methods such as using k-means and HMM [1], or k-NN TSC, showing an improvement of at least 20% in accuracy.

Acknowledgements. We acknowledge the ERC (Project PULSE: ERC-ADG-2015 694581) and the NIHR Oxford Biomedical Research Centre.

References

1. Ahmidi, N., Hager, G.D., Ishii, L., Fichtinger, G., Gallia, G.L., Ishii, M.: Surgical task and skill classification from eye tracking and tool motion in minimally invasive surgery. In: Jiang, T., Navab, N., Pluim, J.P.W., Viergever, M.A. (eds.) MICCAI 2010. LNCS, vol. 6363, pp. 295–302. Springer, Heidelberg (2010). https://doi.org/10.1007/978-3-642-15711-0_37
2. Cai, Y., et al.: Spatio-temporal visual attention modelling of standard biometry plane-finding navigation. Med. Image Anal. **65** (2020). https://doi.org/10.1016/j.media.2020.101762
3. Cho, K., et al.: Learning phrase representations using RNN encoder-decoder for statistical machine translation. In: EMNLP 2014–2014 Conference on Empirical Methods in Natural Language Processing, Proceedings of the Conference, pp. 1724–1734 (2014). https://doi.org/10.3115/v1/d14-1179
4. Droste, R., Cai, Y., Sharma, H., Chatelain, P., Papageorghiou, A.T., Noble, J.A.: Towards capturing sonographic experience: cognition-inspired ultrasound video saliency prediction. In: Zheng, Y., Williams, B.M., Chen, K. (eds.) MIUA 2019. CCIS, vol. 1065, pp. 174–186. Springer, Cham (2020). https://doi.org/10.1007/978-3-030-39343-4_15
5. Droste, R., Chatelain, P., Drukker, L., Sharma, H., Papageorghiou, A.T., Noble, J.A.: Discovering salient anatomical landmarks by predicting human gaze. In: Proceedings - International Symposium on Biomedical Imaging 2020-April, pp. 1711–1714 (2020). https://doi.org/10.1109/ISBI45749.2020.9098505
6. Drukker, L., et al.: Transforming obstetric ultrasound into data science using eye tracking, voice recording, transducer motion and ultrasound video. Sci. Rep. **11**(1), 14109 (2021). https://doi.org/10.1038/s41598-021-92829-1
7. Ebeid, I.A., Bhattacharya, N., Gwizdka, J., Sarkar, A.: Analyzing gaze transition behavior using Bayesian mixed effects Markov models. In: Eye Tracking Research and Applications Symposium (ETRA) (2019). https://doi.org/10.1145/3314111.3319839
8. Fuhl, W., Castner, N., Kübler, T., Lotz, A., Rosenstiel, W., Kasneci, E.: Ferns for area of interest free scanpath classification. In: Eye Tracking Research and Applications Symposium (ETRA) (2019). https://doi.org/10.1145/3314111.3319826
9. Hild, J., Kühnle, C., Voit, M., Beyerer, J.: Predicting observer's task from eye movement patterns during motion image analysis. In: Eye Tracking Research and Applications Symposium (ETRA) (2018). https://doi.org/10.1145/3204493.3204575
10. Lee, Y.H., Wei, C.P., Cheng, T.H., Yang, C.T.: Nearest-neighbor-based approach to time-series classification. Decis. Support Syst. **53**(1), 207–217 (2012). https://doi.org/10.1016/j.dss.2011.12.014. https://www.sciencedirect.com/science/article/pii/S0167923612000097
11. Li, L., et al.: Massively parallel hyperparameter tuning. CoRR abs/1810.0 (2018). http://arxiv.org/abs/1810.05934

12. Liaw, R., Liang, E., Nishihara, R., Moritz, P., Gonzalez, J.E., Stoica, I.: Tune: a research platform for distributed model selection and training (2018). https://arxiv.org/abs/1807.05118

13. Lin, T.Y., Goyal, P., Girshick, R., He, K., Dollár, P.: Focal loss for dense object detection (2018). https://arxiv.org/abs/1708.02002

14. Openvinotoolkit: openvinotoolkit/cvat. https://github.com/openvinotoolkit/cvat

15. Public Health England (PHE): NHS Fetal Anomaly Screening Programme Handbook, August 2018. https://www.gov.uk/government/publications/fetal-anomaly-screening-programme-handbook/20-week-screening-scan

16. Sharma, H., Droste, R., Chatelain, P., Drukker, L., Papageorghiou, A.T., Noble, J.A.: Spatio-temporal partitioning and description of full-length routine fetal anomaly ultrasound scans. In: 2019 IEEE 16th International Symposium on Biomedical Imaging (ISBI 2019), pp. 987–990 (2019). https://doi.org/10.1109/ISBI.2019.8759149

17. Sharma, H., Drukker, L., Chatelain, P., Droste, R., Papageorghiou, A.T., Noble, J.A.: Knowledge representation and learning of operator clinical workflow from full-length routine fetal ultrasound scan videos. Med. Image Anal. **69**, 101973 (2021). https://doi.org/10.1016/j.media.2021.101973. http://www.sciencedirect.com/science/article/pii/S1361841521000190

18. Sharma, H., Drukker, L., Papageorghiou, A.T., Noble, J.A.: Multi-modal learning from video, eye tracking, and pupillometry for operator skill characterization in clinical fetal ultrasound. In: 2021 IEEE 18th International Symposium on Biomedical Imaging (ISBI), pp. 1646–1649 (2021). https://doi.org/10.1109/ISBI48211.2021.9433863

19. Thorndike, R.L.: Who belongs in the family? Psychometrika **18**(4), 267–276 (1953). https://doi.org/10.1007/bf02289263

20. Yamak, P.T., Yujian, L., Gadosey, P.K.: A comparison between ARIMA, LSTM, and GRU for time series forecasting. In: ACM International Conference Proceeding Series, pp. 49–55 (2019). https://doi.org/10.1145/3377713.3377722

Efficient Echocardiogram View Classification with Sampling-Free Uncertainty Estimation

Ang Nan Gu[1(✉)], Christina Luong[1,2], Mohammad H. Jafari[1],
Nathan Van Woudenberg[1], Hany Girgis[2], Purang Abolmaesumi[1],
and Teresa Tsang[1,2]

[1] University of British Columbia, Vancouver, BC, Canada
{guangnan,purang}@ece.ubc.ca
[2] Vancouver General Hospital, Vancouver, BC, Canada

Abstract. View classification is a key initial step for the analysis of echocardiograms. Typical deep learning classifiers can make highly confident errors unnoticed by human operators, consequential for downstream tasks. Instead of failing, it is important to create a method that alarms "I don't know" to inform clinicians of potential errors when faced with difficult or novel inputs. This paper proposes Efficient-Evidential Network (Efficient-EvidNet), a lightweight framework designed to classify echocardiogram views and simultaneously provide a sampling-free uncertainty prediction. Evidential uncertainty is used to filter faulty results and flag out the outliers, hence, improving the overall performance. Efficient-EvidNet classifies among 13 standard echo views with 91.9% test accuracy, competitive with other state-of-the-art lightweight networks. Notably, it achieves a 97.6% test accuracy when only reporting on data with low evidential uncertainty. Further, we propose improved techniques for outlier detection, reaching a 0.97 area under the ROC curve for differentiating between cardiac and lung ultrasound, for which the latter is unseen throughout the training. Efficient-EvidNet does not require costly sampling steps for uncertainty estimation and uses a low parameter neural network, providing two key features that are essential for real-time deployment in clinical scenarios.

Keywords: Uncertainty estimation · Evidential deep learning · View classification · Echocardiography

1 Introduction

Trans-thoracic echocardiography (echo) is a commonly used imaging modality to examine cardiac function and a key to many diagnostic procedures. The

Supported by Canada's NSERC and CIHR.
A. N. Gu and C. Luong—Joint first authors.
P. Abolmaesumi and T. Tsang—Joint senior authors.

emergence of new low-cost portable transducers allows ultrasound in point-of-care (POCUS) settings, markedly improving the accessibility of care. However, the training required to use POCUS effectively is a barrier to broader adoption. Thus, there is a significant demand for automating echo image acquisition and analysis. Deep learning (DL) methods can automate important examination components, such as segmentation, diagnosis of common cardiovascular diseases, and ejection fraction estimation [3, 5, 8, 10].

Designing such automated systems is challenging since clinical labels can be uncertain: echo image acquisition is highly operator dependent, commonly resulting in out-of-distribution (OOD) data points. While experienced clinicians can assess the confidence of their decisions and choose not to diagnose when input is ambiguous, most existing DL solutions for automatic echo analysis lack such capability. Additionally, high prediction confidence does not necessarily imply correctness [13], which can cause critical errors to go unnoticed during patient care. Uncertainty estimation techniques allow algorithms to say "I don't know" in ambiguous contexts. It is specifically crucial for the POCUS environment which presents more ambiguous and/or OOD inputs due to echo acquisition by more novice operators and varied scanning settings.

2   Related Works and Limitations

Echo View Classification: Echo can be acquired from several standard views. View classification is typically a pre-processing step prior to further diagnosis in works such as Stanford's seminal EchoNet [5]. This study focuses on the 13 most widely captured standard echo views illustrated in Fig. 1. Echo view classification algorithms often focus on using Convolutional Neural Networks (CNNs), CNN+LSTM, 3D convolutions, and optical flow [4, 7, 14, 20]. While accurate, these approaches use large models and have long inference times. Recently, lightweight networks for view classification are presented for reducing the inference time and storage requirements [17, 24]. The goal of these works is to enable real-time deployment of computer-assisted echo analysis. However, to the best of our knowledge, all the previous works focus on classification, ignoring the task's inherent uncertainties.

Uncertainty Estimation in Deep Learning: Uncertainty estimation techniques fall under sampling-based and sampling-free categories. Sampling-based methods consider uncertainty as caused by both inherent randomness in data (aleatoric) and the unavailability of data during training (epistemic). They infer with an ensemble of networks [12] or simulate such by using test-time dropout [11, 16], then estimate uncertainty through disagreement of the ensemble. As a result, the inference is lengthy. Sampling-free methods capture uncertainty by estimating data distribution density and exploiting the inherently probabilistic nature of predictions. These include radial basis function networks [23], orthonormal certificates [21], Dirichlet prior networks [15], and evidential neural networks [1, 19]. Not requiring multiple runs of the algorithm to obtain uncertainty improves runtime with the tradeoff typically being the inability to separate aleatoric from epistemic uncertainty.

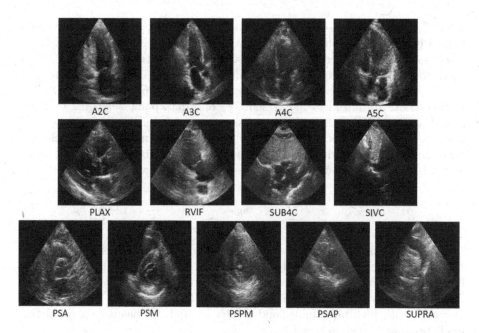

Fig. 1. The standard echo views included in this study.

Our Contributions: Current state-of-the-art (SotA) uncertainty estimation uses sampling with large neural networks. These are costly in time and computational resources, and unsuitable for POCUS deployment. We propose *Efficient-EvidNet*, building on evidential neural networks (ENN) for echo view classification and sampling-free uncertainty quantification, to drastically reduce the number of highly confident yet erroneous predictions. The contributions of this work are as follows: 1) design of Efficient-EvidNet, a lightweight evidential classifier with sampling-free uncertainty; 2) performance evaluation of Efficient-EvidNet; 3) demonstration of robust OOD detection for nonstandard cardiac views and lung ultrasound images; and 4) ablation study of the method to compare against SotA softmax methods for echo view classification.

3   Proposed Method: Efficient-EvidNet

3.1   Background: Evidential Deep Learning

Evidential neural networks for classification predict a Dirichlet distribution over the space of p [19]. The Dirichlet distribution is parameterized by length-K vector α, and is given by:

$$D(p|\alpha) = \begin{cases} \frac{1}{B(\alpha)} \prod_{k=1}^{K} p_k^{\alpha_k - 1}, & \text{if } p \in S_K \\ 0, & \text{otherwise} \end{cases}, \tag{1}$$

where S_K denotes the K-dimensional probability simplex, and $B(\alpha)$ is the multinomial beta function. The typical softmax approach directly estimates p. Using $D(p|\alpha)$ as a second-order distribution instead of direct estimation offers a greater space to model uncertainty. The Dirichlet distribution is an analogue to subjective logic theory where a fundamental notion of uncertainty exists [9]. The parameters α are linked to the evidence e_k of a data point belonging to a certain class. The ENN finds evidence of the data point belonging to each class and updates α. Unlike classical probability, the sum of belief masses b_k for each class is not necessarily one, which allows the uncertainty u to be quantified. α, belief, and uncertainty can be computed from evidence as follows:

$$\alpha_k = e_k + 1, \quad S = \sum_{k=1}^{K} \alpha_k, \quad b_k = \frac{e_k}{S}, \quad u = \frac{K}{S} = 1 - \sum_{k=1}^{K} b_k. \quad (2)$$

The output prediction $\hat{p}_k$ is simply the expected probabilities: $\hat{p}_k = \mathbb{E}[p_k] = \frac{\alpha_k}{S}$. For example, $D(p|\alpha = [1,1,1])$ represents every configuration of $p \in S_K$ being equally probable due to zero evidence, with $b = 0$, $\hat{p}_k = \frac{1}{3}$ and $u = 1$. A distribution with concentrated evidence like $D(p|\alpha = [10,1,1])$ predicts the first class with high probability, and $D(p|\alpha = [10,1,10])$ leads to a conflicting prediction between class 1 and 3.

To train the ENN, we minimize the sum-of-squares loss function between the ground truth y and predicted p (Eq. 3). Intuitively, this translates to simultaneously minimizing the prediction error and Dirichlet variance.

$$\mathcal{L}_i(y_i, p_i|\Theta) = \int \|y_i - p_i\|_2^2 \frac{1}{B(\alpha)} \prod_{j=1}^{K} p_{ij}^{\alpha_{ij}-1} dp_i$$

$$= \sum_{j=1}^{K} (y_{ij}^2 - 2y_{ij}\mathbb{E}[p_{ij}] + \mathbb{E}[p_{ij}^2]) = \sum_{j=1}^{K} (y_{ij} - \mathbb{E}[p_{ij}])^2 + Var[p_{ij}]. (3)$$

3.2   Methodology

Network Architecture: Our approach combines EfficientNet [22], a high performance, low-parameter CNN with the evidential output to produce $D(p|\alpha)$; the proposed methodology is called Efficient-EvidNet (Fig. 2). From $D(p|\alpha)$, it is easy to extract $\hat{p}_k$ and uncertainty. The network is trained on echo images from 13 standard views and some nonstandard views. The uncertainty from Efficient-EvidNet can be used to inform clinicians of the network's confidence, or reject results when the confidence is too low to encourage clinicians to acquire a higher-quality image. The methodology's lightweight and sampling-free nature enables faster feedback to clinicians for prediction and reporting uncertainty.

Training: We present two findings that were helpful in training ENNs. First, class balance in ENNs should be addressed by oversampling classes with less data such that each class appears equally frequently. A loss function weighting approach for class imbalance works for softmax but fails for evidential. Secondly, to

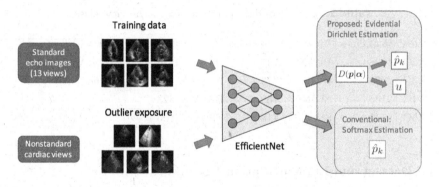

Fig. 2. Block diagram of the proposed approach.

Table 1. Parameter count for lightweight networks used in experiments. MobileNetV1 and Knowledge Distillation are included for comparison with [17,24].

Model	DenseNet-121	EfficientNet-B0	MobileNetV2	MobileNetV1	Knw. Distill.
# of params	7.9M	5.3M	3.4M	4.2M	0.4M

improve OOD detection capability we use an approach inspired by Outlier Exposure (OE) [6], where nonstandard view images are used as exemplar outliers. The network is trained to produce $D(p|[1,..,1])$ for OOD examples, a natural outcome stemming from lack of evidence for any given class. Setting the second-order distribution to be uniform is more intuitive compared to the original OE approach, which demands $\hat{p}_k = \frac{1}{K} \ \forall \ k$ when there is no evidence of the outlier for any class. Equation (4) denotes total loss with $\mathcal{D}$ and $\mathcal{D}_{OE}$ being the in-distribution and outlier exposure data, respectively, and D_{KL} denoting Kullback-Leibler divergence. Regularization λ_r is added to reduce excess evidence predictions; the term $\tilde{\boldsymbol{\alpha}}_i = \boldsymbol{y}_i + (1-\boldsymbol{y}_i) \odot \boldsymbol{\alpha}_i$ denotes the Dirichlet after evidence relating to the majority class is removed. We use $\lambda_r = 0.1$ as done in [19], and $\lambda_{OE} = 1$ but with weighted sampling such that $|\mathcal{D}_{OE}| = \frac{1}{13}|\mathcal{D}|$ to be in proportion with other classes:

$$\mathcal{L}(\Theta) = \sum_{i=1,i\in\mathcal{D}}^{N} \mathcal{L}_i(\boldsymbol{y}_i, \boldsymbol{p}_i|\Theta) + \lambda_r D_{KL}[D(\boldsymbol{p}_i|\tilde{\boldsymbol{\alpha}}_i)||D(\boldsymbol{p}_i|[1,..,1])]$$

$$+\lambda_{OE} \sum_{j=1,j\in\mathcal{D}_{OE}}^{M} D_{KL}[D(\boldsymbol{p}_j|\boldsymbol{\alpha}_j)||D(\boldsymbol{p}_j|[1,..,1])]. \quad (4)$$

4   Experiment

We compare performance of Efficient-EvidNet with three SotA lightweight network structures (DenseNet-121, EfficientNet-B0, and MobileNetV2), and investigate the effect of using the evidential function and outlier exposure technique. The number of parameters can be found in Table 1.

Table 2. Number of data points for each echo view class.

View	# of cines	View	# of cines	View	# of cines	View	# of cines
A2C	1928	PLAX	2747	PSA	2126	SUPRA	76
A3C	2095	RVIF	373	PSM	2265	Outlier	238
A4C	2166	SUB4C	759	PSPM	823		
A5C	541	SIVC	718	PSAP	106		

4.1 Dataset and Implementation Details

Dataset: Echo data used in this paper are agglomerated from several studies at Vancouver General Hospital with approval from the Information Privacy Office and Clinical Medical Research Ethics Board. The data are extracted randomly from the hospital picture archiving system and contain 16612 cines (videos) averaging 48 frames in length each from 3151 unique patients diagnosed with various heart diseases. The dataset contains cines acquired by six devices: *GE Vivid 7, Vivid i, Vivid E9, Philips iE33, Sonosite* and *Sequoia*. An experienced cardiologist examined and labelled the echo view of acquired cines belonging to the 13 different views (Fig. 1). The view distribution is shown in Table 2. The data are split into training/validation/test sets with ratio of 70%/20%/10%, based on mutually exclusive patients. We pose the view classification task as predicting the view of single frames from the cines. This is more challenging compared to other approaches that classify videos. However, here we need to constrain the problem to image-level to resemble real-time deployment in POCUS with limited hardware resources. To test OOD detection, we use a public lung ultrasound (US) image dataset as OOD data. The lung US dataset is compiled by Born et al. [2] for the purpose of COVID-19 detection; it contains lung images from varying angles of anonymous patients with either COVID-19, pneumonia, or no symptoms.

Implementation Details: We use the Adam optimizer with $\beta = (0.9, 0.999)$ and $lr = 2.5e-4$. We train networks for 30 epochs and exponential $lr\ decay = 0.95$, initializing from pre-trained *ImageNet* weights. For training data augmentation, the images are resized to 256×256 and rotated randomly by $\pm 5°$. Then, a 224×224 region is cropped randomly to match the network input dimension. For validation and testing, the image is simply resized to 224×224 pixels.

Metrics: For each experiment we measure the classifier accuracy and the OOD detection performance of the uncertainty metric, a scalar in $[0, 1]$. For the softmax approach, the metric is confidence $\hat{p}_k$ of the chosen class, for evidential the metric is u from Eq. (2). OOD AUC is the area under the receiver operating characteristic curve (AUROC) of classifying standard cardiac views versus nonstandard views and lung ultrasound using the uncertainty as a threshold.

Table 3. Results for ablation experiments and comparison to previous works featuring lightweight networks. Best results for each category are in bold.

Experiment	Accuracy	OOD AUC	Experiment	Accuracy	OOD AUC
MobileNet+SM	90.8	90.4	DenseNet+SM	90.4	93.7
Mobile+SM+OE	89.7	95.7	DenseNet+SM+OE	89.8	96.6
MobileNet+Evid	89.1	87.0	DenseNet+Evid	87.6	82.2
Mobile+Evid+OE	85.5	96.9	DenseNet+Evid+OE	87.9	96.9
Knw. Distill	89.0	–	EffNet+SM	91.4	90.1
MobileNetV1	88.1	–	EffNet+SM+OE	89.3	*97.0*
			EffNet+Evid (Proposed)	*91.9*	91.1
			EffNet+Evid+OE	89.4	*97.0*

4.2   Results and Discussion

Functionality of the Evidential Output. For most cases, the accuracy of softmax (denoted SM in Table 3) is competitive with evidential without reject-ing uncertain samples (Table 3). The rejection-based accuracy diverges as more data points are filtered (Fig. 3). Evidential uncertainty is more expressive at the top-end since the asymptotic relationship between u and evidence saturates slower compared to the exponential relationship between softmax confidence p_k and logits. The increased expressiveness of evidential uncertainty leads to more complete characterization of the error/rejection rate curve compared to softmax, where a majority of predictions for test data exceed the 0.98 confidence mark.

Choosing the Uncertainty Rejection Threshold. In practice, we need to define a point on the rejection/accuracy curve for the model to operate. For the softmax approach, the rejection point is where confidence is lower than 0.98. For evidential the rejection point is chosen by selecting the optimum threshold on the validation set using knee point detection [18] (Fig. 3). For most ablations, the evidential elbow point is located at a higher rejection rate than softmax. Figure 5 shows performance for each class before and after accounting for rejection.

When is the Model Uncertain? Many parasternal window images report higher uncertainties. This could be due to a higher degree of geometric similar-ity between the mid-level parasternal views (PSM/PSPM/PSAP). Images with lower visual resolution where anatomical features are shown less prominently, or images with ambiguous visual geometry (such as A4C displayed in Fig. 4) have higher uncertainty. Most of the lung ultrasound images are correctly flagged as outliers based on uncertainty. For both softmax and evidential, adding out-lier exposure slightly decreases the baseline test accuracy, but greatly boosts the OOD detection capability. Overall, Efficient-EvidNet is proposed for its high accuracy at baseline and with rejection. Efficient-EvidNet+OE is suggested when stronger OOD detection is required.

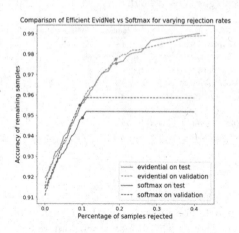

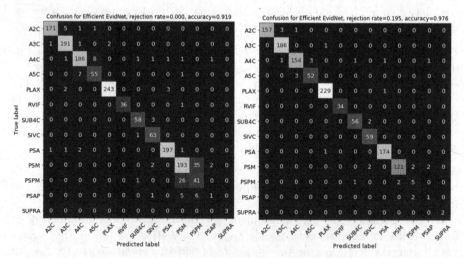

Fig. 3. Validation/test accuracy plotted for 100 uncertainty thresholds evenly spaced between [0, 0.99]. Dots indicate the threshold selected on the validation set, then applied on the test set.

Fig. 4. Visual samples of low uncertainty (left column) vs high uncertainty (right column) images.

Fig. 5. View classification confusion matrix at the baseline accuracy (left) and rejection point accuracy (right).

5   Conclusion

Evidential deep learning can be a powerful alternative to softmax, providing sampling-free uncertainty quantification and outlier detection, both of which are very helpful in real-life data acquisition scenarios. We adopt the principles of evidential learning to several SotA architectures and propose Efficient-EvidNet, using EfficientNet-B0 as a backbone with outlier exposure-augmented training. Efficient-EvidNet reaches 97.6% accuracy in echo view detection by automatically detecting and rejecting 19.5% of data with the highest evidential uncertainty. Future work includes extension of the proposed methodology to downstream tasks such as segmentation and automated cardiac diagnosis.

References

1. Amini, A., Schwarting, W., Soleimany, A., Rus, D.: Deep evidential regression. In: Larochelle, H., Ranzato, M., Hadsell, R., Balcan, M.F., Lin, H. (eds.) Advances in Neural Information Processing Systems, vol. 33, pp. 14927–14937. Curran Associates, Inc. (2020)
2. Born, J., et al.: Accelerating detection of lung pathologies with explainable ultrasound image analysis. Appl. Sci. **11**(2), 672 (2021). https://doi.org/10.3390/app11020672
3. Chen, C., et al.: Deep learning for cardiac image segmentation: a review. Front. Cardiovasc. Med. **7**, 25 (2020). https://doi.org/10.3389/fcvm.2020.00025. https://www.frontiersin.org/article/10.3389/fcvm.2020.00025
4. Gao, X., Li, W., Loomes, M., Wang, L.: A fused deep learning architecture for viewpoint classification of echocardiography. Inf. Fusion **36**, 103–113 (2017). https://doi.org/https://doi.org/10.1016/j.inffus.2016.11.007. https://www.sciencedirect.com/science/article/pii/S1566253516301385
5. Ghorbani, A., et al.: Deep learning interpretation of echocardiograms. bioRxiv (2019). https://doi.org/10.1101/681676. https://www.biorxiv.org/content/early/2019/06/24/681676
6. Hendrycks, D., Mazeika, M., Dietterich, T.: Deep anomaly detection with outlier exposure. arXiv preprint arXiv:1812.04606 (2018)
7. Howard, J., et al.: Improving ultrasound video classification: an evaluation of novel deep learning methods in echocardiography. J. Med. Artif. Intell. **3**, 4 (2020). https://doi.org/10.21037/jmai.2019.10.03
8. Jafari, M.H., et al.: Automatic biplane left ventricular ejection fraction estimation with mobile point-of-care ultrasound using multi-task learning and adversarial training. Int. J. Comput. Assist. Radiol. Surg. **14**(6), 1027–1037 (2019). https://doi.org/10.1007/s11548-019-01954-w
9. Jøsang, A.: Subjective Logic. Springer, Cham (2016). https://doi.org/10.1007/978-3-319-42337-1
10. Kazemi Esfeh, M.M., Luong, C., Behnami, D., Tsang, T., Abolmaesumi, P.: A deep Bayesian video analysis framework: towards a more robust estimation of ejection fraction. In: Martel, A.L., et al. (eds.) MICCAI 2020. LNCS, vol. 12262, pp. 582–590. Springer, Cham (2020). https://doi.org/10.1007/978-3-030-59713-9_56
11. Kendall, A., Gal, Y.: What uncertainties do we need in Bayesian deep learning for computer vision? arXiv preprint arXiv:1703.04977 (2017)
12. Lakshminarayanan, B., Pritzel, A., Blundell, C.: Simple and scalable predictive uncertainty estimation using deep ensembles. arXiv preprint arXiv:1612.01474 (2016)
13. Li, Z., Hoiem, D.: Improving confidence estimates for unfamiliar examples. In: Proceedings of the IEEE/CVF Conference on Computer Vision and Pattern Recognition, pp. 2686–2695 (2020)

14. Madani, A., Arnaout, R., Mofrad, M., Arnaout, R.: Fast and accurate view classification of echocardiograms using deep learning. NPJ Digit. Med. **1**(1), 1–8 (2018)
15. Malinin, A., Gales, M.: Predictive uncertainty estimation via prior networks. arXiv preprint arXiv:1802.10501 (2018)
16. Moshkov, N., Mathe, B., Kertesz-Farkas, A., Hollandi, R., Horvath, P.: Test-time augmentation for deep learning-based cell segmentation on microscopy images. Sci. Rep. **10**(1), 1–7 (2020)
17. Pop, D.: Classification of heart views in ultrasound images. Master's thesis, Linköping University, Computer Vision (2020)
18. Satopaa, V., Albrecht, J., Irwin, D., Raghavan, B.: Finding a "kneedle" in a haystack: Detecting knee points in system behavior. In: 2011 31st International Conference on Distributed Computing Systems Workshops, pp. 166–171. IEEE (2011)
19. Sensoy, M., Kaplan, L., Kandemir, M.: Evidential deep learning to quantify classification uncertainty. arXiv preprint arXiv:1806.01768 (2018)
20. Shahin, A.I., Almotairi, S.: An accurate and fast cardio-views classification system based on fused deep features and LSTM. IEEE Access **8**, 135184–135194 (2020). https://doi.org/10.1109/ACCESS.2020.3010326
21. Tagasovska, N., Lopez-Paz, D.: Single-model uncertainties for deep learning. arXiv preprint arXiv:1811.00908 (2018)
22. Tan, M., Le, Q.: EfficientNet: rethinking model scaling for convolutional neural networks. In: International Conference on Machine Learning, pp. 6105–6114. PMLR (2019)
23. Van Amersfoort, J., Smith, L., Teh, Y.W., Gal, Y.: Uncertainty estimation using a single deep deterministic neural network. In: International Conference on Machine Learning, pp. 9690–9700. PMLR (2020)
24. Vaseli, H., et al.: Designing lightweight deep learning models for echocardiography view classification. In: Medical Imaging 2019: Image-Guided Procedures, Robotic Interventions, and Modeling, vol. 10951, p. 109510F. International Society for Optics and Photonics (2019)

Contrastive Learning for View Classification of Echocardiograms

Agisilaos Chartsias[1]([⊠]), Shan Gao[1], Angela Mumith[1], Jorge Oliveira[1], Kanwal Bhatia[1,2], Bernhard Kainz[1,3], and Arian Beqiri[1,4]

[1] Ultromics Ltd., 4630 Kingsgate, Cascade Way, Oxford Business Park South, Oxford OX4 2SU, UK
`agis.chartsias@ultromics.com`
[2] Metalynx Ltd., 71-75 Shelton Street, London WC2H 9JQ, UK
[3] Imperial College London, 180 Queen's Gate, London SW7 2AZ, UK
[4] School of Biomedical Engineering and Imaging Sciences, King's College London, London SE1 7EU, UK

Abstract. Analysis of cardiac ultrasound images is commonly performed in routine clinical practice for quantification of cardiac function. Its increasing automation frequently employs deep learning networks that are trained to predict disease or detect image features. However, such models are extremely data-hungry and training requires labelling of many thousands of images by experienced clinicians. Here we propose the use of contrastive learning to mitigate the labelling bottleneck. We train view classification models for imbalanced cardiac ultrasound datasets and show improved performance for views/classes for which minimal labelled data is available. Compared to a naïve baseline model, we achieve an improvement in F1 score of up to 26% in those views while maintaining state-of-the-art performance for the views with sufficiently many labelled training observations.

Keywords: Contrastive learning · Classification · Echocardiography

1 Introduction

Echocardiography is widely and routinely used for assessing heart function and for the diagnosis of several conditions, such as heart failure and coronary artery disease [13]. In a routine echocardiographic study, multiple views of the heart are obtained to show different parts of the heart's internal structure, i.e. the ventricles, atria and valves—see Fig. 1. However, not all views are used in subsequent analysis of the echocardiograms depending on the cardiac function being assessed or the type of disease being investigated [13]. Therefore, an important initial step in any automated analysis pipeline is the accurate detection of standardised cardiac views shown on each echocardiogram. Frequently, further analysis—usually performed with proprietary analysis software—focuses on left ventricular function [17]. Often only the three apical views of the heart are assessed, which show

© Springer Nature Switzerland AG 2021
J. A. Noble et al. (Eds.): ASMUS 2021, LNCS 12967, pp. 149–158, 2021.
https://doi.org/10.1007/978-3-030-87583-1_15

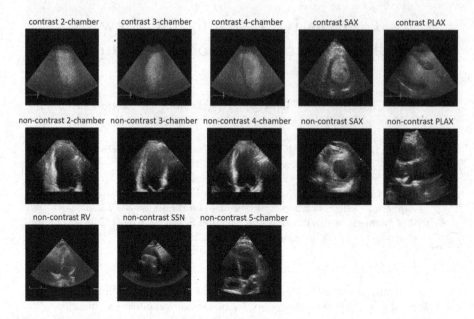

Fig. 1. Examples of different echocardiography views used including the 2/3/4/5 chamber apical, parasternal long-axis (PLAX), short-axis (SAX) at papillary muscle, right ventricular (RV) and suprasternal notch (SSN) views. The top row shows images obtained after injection of a microbubble contrast agent, causing a near inversion in image contrast, whereas the lower two rows show non-contrast images.

slices through the left ventricle. However, it is still important for a view classifier to be aware of the entire cardiac anatomy so that it does not misclassify views it has not been trained on. This is challenging because it requires large training datasets with appropriate labels. Furthermore, when assessing certain cardiac conditions, the injection of a microbubble contrast agent is used to better highlight the boundaries of the left ventricular wall [20]. This changes the image appearance completely and effectively inverts the image contrast. Hence, these views cannot be classified without contrast enhanced data also being labelled for model training. The ability to correctly classify contrast images thus requires double the labelling effort.

View classification on echocardiographic data has previously been achieved using convolutional networks [8,18,22] that take as input an image and predict one of the possible views that were present in the training label set for that network. For the commonly acquired echocardiographic views, such as the apical four-chamber view, labelled data for model training is available even in some public datasets [14,19]. However, for less commonly acquired views, with or without contrast enhancement, it is time-consuming and expensive to acquire labels and thus, datasets are often highly imbalanced. To tackle data imbalance, training classifiers may require under-sampling the majority classes and specialised cost functions [10] or augmentations with synthetically generated data [1].

In this paper, we investigate the problem of view classification in cardiac ultrasound images and attempt to improve the classification accuracy of convolutional neural networks, especially on under-represented classes, with the use of contrastive learning. Contrastive learning is a pre-training methodology, which improves learning of features useful for classification tasks through a contrastive loss. The contrastive loss clusters similar images together (positive pairs) and pushes different images away (negative pairs). This can be entirely based on self supervision for example when positive pairs consist of different augmented version of an image (SimCLR [6]) or, when in addition to augmentations, positive pairs also use supervision to include images of the same label (SupCon [11]). This has proven successful in computer vision tasks for instance for ImageNet sample classification [6].

Furthermore, although cardiac ultrasound data consist of videos, view classification is typically performed per-frame as a 2D classification problem. For videos, unsupervised contrastive learning, such as SimCLR, is not directly applicable as also discussed in [7]: if multiple frames of the same video end up in the same batch, then the negative pairs of a frame will include other frames of the same video. This would hinder the ability of the contrastive loss to only cluster similar images together, since different video frames would generate a higher loss value. We therefore adopt the supervised contrastive loss [11], which does not suffer from this limitation. Our contributions are the following: (a) we apply contrastive classification neural networks to cardiac ultrasound, and (b) we evaluate in a dataset of contrast and non-contrast enhanced echocardiographic images collated from public and proprietary sources and show improved results when using the proposed contrastive framework for views which have fewer labelled training observations.

2   Related Work

Standard plane/view detection has been previously studied in fetal ultrasound with supervised deep learning models, such as SonoNet [2], multi-scale DenseNet [12], and convolutional networks finetuned with transfer learning [5] or trained with additional tasks to predict attention maps and adversarial training [3]. In echocardiography, inception [18] and VGG [22] networks have been used to predict several views or subclasses of views, although not applied on contrast echo data. Typically, contrast-enhanced images are used in isolation, for example to extract myocardial segmententations [15,16]. Most recently, high view classification accuracy was reported by a convolutional network applied on mixed microbubble contrast-enhanced and non-contrast data from a multi-vendor site [8].

Given sufficiently large datasets, supervised training of convolutional networks is successful in accurate view detection. However, network initialisation is important to facilitate convergence, and therefore pre-training methods using self-supervision with different augmented views of the same image [6] or labels [11] are investigated to improve computer vision classification tasks, such

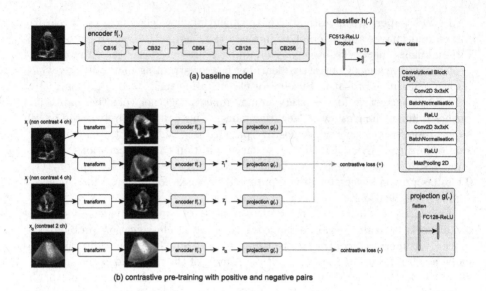

Fig. 2. Schematic of the baseline and contrastive models. (a) The baseline model architecture consists of a fully convolutional encoder and a fully connected classifier, and is trained with full supervision. (b) The contrastive model pre-trains the encoder using a projection network and a contrastive loss. The contrastive loss considers positive pairs if these are different augmentations of the same image or belong to the same class, and negative pairs otherwise.

as on the ImageNet dataset. Contrastive learning has also been used in the medical domain, for instance to improve segmentation performance on MRI images [4] or to learn joint representations of ultrasound videos and speech [9].

3   Methodology

Given an image x of view y_k, where $k \in [1, 13]$, corresponding to 13 classes of commonly acquired views with or without contrast, we consider a 2D baseline classification neural network $c(x)$ to detect per-frame view labels. This network maps input images through five convolutional blocks, each containing two convolutional layers followed by batch normalisation and a ReLU activation function, and a max pooling layer, to a vector representation, which is then processed by two fully connected layers to generate a view label prediction. This model architecture, which was used in an eight-class form in [8], is designed so that it is sufficiently small and effective on standard view classification and can be seen in Fig. 2a. Training is performed with the categorical cross entropy loss described as follows:

$$L_{view} = -\sum_{k=1}^{13} y_k log(c(x)).$$

A contrastive learning framework is then implemented as per the SupCon [11] methodology as follows: we split the baseline model into a fully convolutional and a fully connected sub-model, which are used as an encoder $f(.)$ and classification network $h(.)$, respectively so that $c = h \circ f$. We add a projection network $g(.)$, which projects the encoded features $z = f(x)$ into a representation $\hat{x} = g(z)$. The projection $\hat{x}$ is used as an input to the contrastive loss that pre-trains the encoder. Finally, the classification network $h(.)$ learns a mapping of the encoded features to their corresponding labels and is trained on a second stage following the encoder pre-training, whilst keeping the encoder weights fixed. A schematic of the framework is shown in Fig. 2.

The contrastive learning process is more formally described as: given N randomly augmented images $\{x_i\}_{i=1}^N$, we first obtain a batch of $2N$ images $B = \{1 \ldots 2N\}$ by applying a second augmentation. For every image x_i in the batch, and its projection $\hat{x}_i = g(f(x_i))$, there are also M_i other images of the same label in the set $P_i = \{x_j\}_{j=1}^{M_i}$. According to [11] the supervised contrastive loss is defined as:

$$L_{supcon} = \sum_{i \in B} -log \left\{ \frac{1}{M_i} \sum_{j \in P_i} \frac{exp(\hat{x}_i \hat{x}_j / \tau)}{\sum_{\alpha \in B \backslash i} exp(\hat{x}_i \hat{x}_\alpha / \tau)} \right\},$$

where τ controls the temperature scaling of the softmax. We set $\tau = 1000$ as per [11] and use brightness and contrast augmentations, as well as rotations to $30°$ and spatial translations at up to 10% of the image dimensions.

3.1 Data

The dataset used in this work comprised of anonymised 2D echocardiograms from multiple sites. The dataset is composed of data from EVAREST [21], a multi-site, multi-vendor UK trial, some data from the EchoNet public dataset [19],

Table 1. Description of the training and test dataset.

Contrast	View	Training set			Test set		
		Subjects	Echocardiograms	Images	Subjects	Echocardiograms	Images
✓	2 ch.	711	1401	41603	139	276	5784
✓	3 ch.	699	1377	41432	139	276	5763
✓	4 ch.	704	1375	41214	138	274	5684
✓	plax	85	85	4547	9	9	560
✓	sax	607	1179	34387	138	275	5649
✗	5 ch.	165	165	14714	18	18	1832
✗	plax	383	383	33969	42	42	3542
✗	rv	52	52	6521	5	5	703
✗	ssn	55	55	3483	6	6	336
✗	2 ch.	314	544	26613	126	217	7061
✗	3 ch.	364	605	32263	135	226	8662
✗	4 ch.	332	556	28569	130	205	7938
✗	sax	229	437	17704	98	187	4234

and some proprietary data from other imaging sites. The final dataset is split into a training and a test set of echocardiograms corresponding to 1,538 and 359 subjects, respectively. The total number of image frames contained in these data is 327,019 for the training set and 57,648 for the test set. Each echocardiographic video was labelled into one of 13 classes, which cover a set of standard cardiac views with or without microbubble contrast. The classes are shown in the first and second columns of Table 1 along with the number of subjects, echocardiograms and images present for each view.

Images were extracted from DICOM or AVI files and were pre-processed to remove all text information and annotations outside the ultrasound sector, so that the dataset contains only the images within the ultrasound sector.

As part of the EVAREST trial data, the dataset contains echocardiograms obtained with the patient at rest and with patients subjected to exercise or pharmacological stress. Heart rates vary from 45 to 150 and the number of heartbeats per scan are between one and three. The inclusion of stress echo data ensures that a range of image qualities is present in the dataset as stress echocardiograms tend to include images of poor image quality.

4   Experiments and Discussion

4.1   Experimental Setup

Prior to being fed into the network, image frames are resized to 192×192 pixel size, z-score normalised, and rescaled to $[0, 1]$ range. The model and pipeline was developed in Python 3.7.7 with Tensorflow 2.2 and training was performed on four Nvidia GeForce RTX 2080 Ti GPUs with 11 GB VRAM each.

The baseline and contrastive learning methods were trained using Adam with batch size 64[1] and learning rate equal to 0.0001 on a 8-fold cross-validation with the validation set containing 10% of the training dataset's echocardiograms. Training stopped using an early stopping criterion based on the validation set.

We train models using all 13 view classes in two scenarios: one using all data, and then one with reduced data of around 50 echocardiograms per class, chosen at random. We report the mean F1 score, precision and recall across the different validation splits and a held out test set that is common across the different splits.

4.2   Classification Performance

Table 2 shows the mean and standard deviation of F1 score, precision and recall for the experiments on the full and reduced datasets. Both methods perform equally well on the dataset of 50 echocardiograms per class, which is balanced. We observe an improvement in test F1 score on the full dataset, which increases from 0.874 to 0.892, and smaller standard deviations in precision and recall.

Table 3 reports the per-class test F1 score for the two datasets. When assessing the per-class classifier performance, it can be seen that the contrastive

---

[1] The effective batch size is 128, since every image is augmented twice in a batch.

Table 2. Classification results (mean and standard deviation) of baseline and contrastive models on validation (taken from 10% of the training set) and test sets using two datasets containing all data and 50 echocardiograms per class, respectively.

Dataset	Method	Validation set			Test set		
		F1 Score	Precision	Recall	F1 Score	Precision	Recall
50 echocardiograms	Baseline	$0.794_{.02}$	$0.780_{.02}$	$0.837_{.01}$	$0.765_{.02}$	$0.756_{.03}$	$0.820_{.02}$
per class	SupCon	$0.800_{.01}$	$0.787_{.02}$	$0.833_{.01}$	$0.775_{.01}$	$0.770_{.02}$	$0.825_{.01}$
All data	Baseline	$0.911_{.02}$	$0.924_{.03}$	$0.902_{.02}$	$0.874_{.01}$	$0.896_{.02}$	$0.880_{.02}$
	SupCon	$0.915_{.02}$	$0.928_{.01}$	$0.908_{.02}$	$0.892_{.01}$	$0.907_{.01}$	$0.896_{.01}$

Table 3. Classification results (mean and standard deviation) per class. The first column indicates whether the images have contrast or not. Results show the F1 score on the test set for two experiments using different training set sizes, with the number of studies for each view shown. Highest differences are marked in bold.

Cont	View	Size	Baseline	SupCon	%Diff	Size	Baseline	SupCon	%Diff
✓	2 ch.	50	$0.693_{.03}$	$0.702_{.02}$	1.24	677	$0.866_{.01}$	$0.870_{.01}$	0.42
✓	3 ch.	50	$0.811_{.02}$	$0.811_{.03}$	0.02	664	$0.966_{.00}$	$0.968_{.00}$	0.22
✓	4 ch.	50	$0.758_{.07}$	$0.737_{.06}$	−2.73	672	$0.888_{.00}$	$0.896_{.01}$	0.99
✓	plax	50	$0.608_{.16}$	$0.733_{.05}$	**20.61**	68	$0.570_{.08}$	$0.719_{.09}$	**26.08**
✓	sax	50	$0.926_{.04}$	$0.946_{.01}$	2.22	570	$0.985_{.00}$	$0.986_{.00}$	0.12
✗	5 ch.	50	$0.546_{.05}$	$0.542_{.04}$	−0.76	132	$0.660_{.05}$	$0.706_{.05}$	**6.98**
✗	plax	50	$0.952_{.03}$	$0.959_{.02}$	0.83	306	$0.972_{.01}$	$0.974_{.01}$	0.15
✗	rv	42	$0.358_{.06}$	$0.363_{.06}$	1.45	42	$0.632_{.12}$	$0.697_{.09}$	**10.26**
✗	ssn	44	$0.700_{.06}$	$0.679_{.04}$	−3.03	44	$0.990_{.03}$	$0.952_{.04}$	−3.88
✗	2 ch.	50	$0.857_{.01}$	$0.856_{.01}$	−0.04	269	$0.939_{.00}$	$0.934_{.01}$	−0.54
✗	3 ch.	50	$0.912_{.01}$	$0.910_{.01}$	−0.24	319	$0.967_{.01}$	$0.969_{.00}$	0.22
✗	4 ch.	50	$0.879_{.01}$	$0.877_{.01}$	−0.23	287	$0.937_{.01}$	$0.936_{.01}$	−0.06
✗	sax	50	$0.951_{.01}$	$0.963_{.01}$	1.27	213	$0.988_{.00}$	$0.989_{.00}$	0.10

training has minimal effect for the model trained on 50 echocardiograms per class. When training on the full dataset, classes which have a larger number of training data show similar or marginal improvement in performance in the test set. However, classes with substantially less training data, such as the contrast PLAX view, non-contrast 5-chamber view, and the non-contrast right ventricular (RV) view show greater improvement when using contrastive learning. The non-contrast suprasternal notch (SSN) view shows a 4% reduction but both baseline and contrastive model accuracies are very high.

4.3 Ablation Studies and Failure Cases

We perform two ablation experiments on the model parameters. Firstly, we evaluate the effect of batch size by testing values equal to 32 and 16. The obtained results are the same as the ones achieved with batch size 64. Although it has

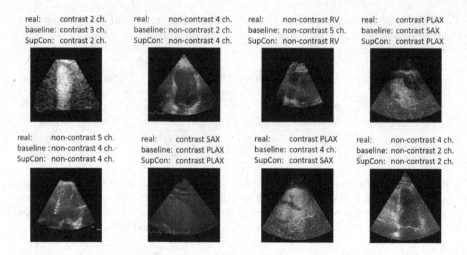

Fig. 3. Selection of failure cases. The baseline model fails on all these, but SupCon correctly classifies the examples in the top row.

been reported that large batch sizes benefit contrastive learning [6], since more positive and negative examples occur in a batch, at this value range the effect is minimal. GPU memory limitations prevented experiments with higher values.

We also experiment with different sets of augmentations. The experiments in Sect. 4.2 use random rotations, translations, as well as changes in brightness and contrast. Random crops resulting in images of 140×140 pixel size have also been tested. However, training with such crop augmentations decreased the validation F1 score of the contrastive model by approximately 15%. This can be attributed to the fact that in view classification, cropped ultrasound images might generate images which appear similar to other views.

Finally, Fig. 3 shows a selection of cases for which the baseline model fails, but for some the contrastive model is able to predict correctly. In all cases, the incorrect view is visually similar to the true view (for example, the apical 4 and 5 chamber views are very similar) so it is evident why the models would struggle. The contrastive model is likely more successful with these challenging views as it creates a better decision boundary between classes.

5   Conclusion

We have shown that the use of contrastive learning applied to echocardiographic view classification can improve accuracy and reduce standard deviation of the classifier for views for which far less training data is available, with no reduction in overall performance. This indicates that contrastive learning could be a powerful tool in developing models for analysing medical images without requiring such intensive collection and labelling of very large datasets.

We leave as future work testing the effect of different contrastive losses on diverse datasets potentially including unlabelled data, as well as studying the effect of design biases introduced by different encoder architectures on the quality of the learnt latent representations.

Acknowledgements. We thank the echocardiographers involved in this study for their thorough annotation of images from the EVAREST dataset.

References

1. Antoniou, A., Storkey, A., Edwards, H.: Data augmentation generative adversarial networks. In: International Conference on Learning Representations Workshop (2018)
2. Baumgartner, C.F., et al.: SonoNet: real-time detection and localisation of fetal standard scan planes in freehand ultrasound. IEEE Trans. Med. Imaging **36**(11), 2204–2215 (2017)
3. Cai, Y., Sharma, H., Chatelain, P., Noble, J.A.: Multi-task SonoEyeNet: detection of fetal standardized planes assisted by generated sonographer attention maps. In: Frangi, A.F., Schnabel, J.A., Davatzikos, C., Alberola-López, C., Fichtinger, G. (eds.) MICCAI 2018. LNCS, vol. 11070, pp. 871–879. Springer, Cham (2018). https://doi.org/10.1007/978-3-030-00928-1_98
4. Chaitanya, K., Erdil, E., Karani, N., Konukoglu, E.: Contrastive learning of global and local features for medical image segmentation with limited annotations. In: Advances in Neural Information Processing Systems, vol. 33 (2020)
5. Chen, H., et al.: Standard plane localization in fetal ultrasound via domain transferred deep neural networks. IEEE J. Biomed. Health Inform. **19**(5), 1627–1636 (2015)
6. Chen, T., Kornblith, S., Norouzi, M., Hinton, G.: A simple framework for contrastive learning of visual representations. In: International Conference on Machine Learning, pp. 1597–1607. PMLR (2020)
7. Chen, Y., et al.: Effective sample pair generation for ultrasound video contrastive representation learning. arXiv preprint arXiv:2011.13066 (2020)
8. Gao, S., et al.: Fully automated contrast and non-contrast cardiac view detection in echocardiography a multi-centre, multi-vendor study. Eur. Heart J. **41**(Supplement_2), ehaa946-0078 (2020)
9. Jiao, J., Cai, Y., Alsharid, M., Drukker, L., Papageorghiou, A.T., Noble, J.A.: Self-supervised contrastive video-speech representation learning for ultrasound. In: Martel, A., et al. (eds.) MICCAI 2020. LNCS, vol. 12263, pp. 534–543. Springer, Cham (2020). https://doi.org/10.1007/978-3-030-59716-0_51
10. Johnson, J.M., Khoshgoftaar, T.M.: Survey on deep learning with class imbalance. J. Big Data **6**(1), 1–54 (2019)
11. Khosla, P., et al..: Supervised contrastive learning. In: Advances in Neural Information Processing Systems, vol. 33 (2020)
12. Kong, P., Ni, D., Chen, S., Li, S., Wang, T., Lei, B.: Automatic and efficient standard plane recognition in fetal ultrasound images via multi-scale dense networks. In: Melbourne, A., et al. (eds.) PIPPI/DATRA -2018. LNCS, vol. 11076, pp. 160–168. Springer, Cham (2018). https://doi.org/10.1007/978-3-030-00807-9_16

13. Lang, R.M., et al.: Recommendations for cardiac chamber quantification by echocardiography in adults: an update from the American society of echocardiography and the European association of cardiovascular imaging. J. Am. Soc. Echocardiogr. **28**, 1-39.e14 (2015)

14. Leclerc, S., et al.: Deep learning for segmentation using an open large-scale dataset in 2D echocardiography. IEEE Trans. Med. Imaging **38**(9), 2198–2210 (2019)

15. Li, M., et al.: A deep learning approach with temporal consistency for automatic myocardial segmentation of quantitative myocardial contrast echocardiography. Int. J. Cardiovasc. Imaging 1–12 (2021)

16. Li, Y., Ho, C.P., Toulemonde, M., Chahal, N., Senior, R., Tang, M.X.: Fully automatic myocardial segmentation of contrast echocardiography sequence using random forests guided by shape model. IEEE Trans. Med. Imaging **37**(5), 1081–1091 (2017)

17. Nagueh, S.F., et al.: Recommendations for the evaluation of left ventricular diastolic function by echocardiography: an update from the American society of echocardiography and the European association of cardiovascular imaging. J. Am. Soc. Echocardiogr. **29**(4), 277–314 (2016)

18. Østvik, A., Smistad, E., Aase, S.A., Haugen, B.O., Lovstakken, L.: Real-time standard view classification in transthoracic echocardiography using convolutional neural networks. Ultrasound Med. Biol. **45**(2), 374–384 (2019)

19. Ouyang, D., et al.: Video-based AI for beat-to-beat assessment of cardiac function. Nature **580**(7802), 252–256 (2020)

20. Pellikka, P.A., et al.: Guidelines for performance, interpretation, and application of stress echocardiography in ischemic heart disease: from the American society of echocardiography. J. Am. Soc. Echocardiogr. **33**(1), 1-41.e8 (2020)

21. Woodward, W., et al.: Real-world performance and accuracy of stress echocardiography: the EVAREST observational multi-centre study. Eur. Heart J. Cardiovasc. Imaging **44**(March), 1–10 (2021)

22. Zhang, J., et al.: Fully automated echocardiogram interpretation in clinical practice: feasibility and diagnostic accuracy. Circulation **138**(16), 1623–1635 (2018)

Imaging Biomarker Knowledge Transfer for Attention-Based Diagnosis of COVID-19 in Lung Ultrasound Videos

Tyler Lum[1]([envelope]), Mobina Mahdavi[1], Oron Frenkel[2], Christopher Lee[3],
Mohammad H. Jafari[1], Fatemeh Taheri Dezaki[1], Nathan Van Woudenberg[1],
Ang Nan Gu[1], Purang Abolmaesumi[1], and Teresa Tsang[3]

[1] Department of Electrical and Computer Engineering,
University of British Columbia, Vancouver, BC, Canada
`tyler.lum@alumni.ubc.ca`
[2] Providence Health Care, Vancouver, BC, Canada
[3] Vancouver General Hospital, Vancouver, BC, Canada

Abstract. The use of lung ultrasound imaging has recently emerged as a quick, cost-effective, and safe method for diagnosis of patients with COVID-19. Challenges with training deep networks to identify COVID-19 signatures in lung ultrasound data are that large datasets do not yet exist; disease signatures are sparse, but are spatially and temporally correlated; and signatures may appear sporadically in ultrasound video sequences. We propose an attention-based video model that is specifically designed to detect these disease signatures, and leverage a knowledge transfer approach to overcome existing limitations in data availability. In our design, a convolutional neural network extracts spatially encoded features, which are fed to a transformer encoder to capture temporal information across the frames and focus on the most important frames. We guide the network to learn clinically relevant features by training it on a pulmonary biomarker detection task, and then transferring the model's knowledge learned from this problem to achieve 80% precision and 87% recall for COVID-19. Our results outperform the state-of-the-art model on a public lung ultrasound dataset. We perform ablation studies to highlight the efficacy of our design over previous state-of-the-art frame-based approaches. To demonstrate that our approach learns clinically relevant imaging biomarkers, we introduce a novel method for generating attention-based video classification explanations called Biomarker Attention-scaled Class Activation Mapping (Bio-AttCAM). Our analysis of the activation map shows high correlation with the key frames selected by clinicians.

Keywords: COVID-19 lung ultrasound · Attention-based video classification · Knowledge transfer · Imaging biomarkers · Explainability

---

T. Lum and M. Mahdavi contributed equally to this work.

© Springer Nature Switzerland AG 2021
J. A. Noble et al. (Eds.): ASMUS 2021, LNCS 12967, pp. 159–168, 2021.
https://doi.org/10.1007/978-3-030-87583-1_16

1   Introduction

The Sars-Cov-2 pandemic has caused over 4 million deaths and has infected over 200 million people worldwide [7]. Fast, early, and accurate diagnosis of COVID-19 infection is crucial for disease management. The standard genetic test (RT-PCR) has a high sensitivity, but suffers from a long 24 h processing time [6]. Serology tests detect the presence antibodies produced in the body, so they are unable to detect the infection in the early stages.

Point-of-Care Ultrasound (POCUS), however, has gained attention in this area, as it avoids many of these challenges. It is cost effective and portable, which allows it to be used in under-resourced settings [5]. It is also very fast to set up and sanitize the device between patients, and does not use ionizing radiation, which makes it ideal for wide-scale use in assisting the diagnosis and management of pulmonary diseases [1]. However, a significant limitation of ultrasound imaging is the amount of expertise required to interpret the images [5]. Deep learning-based ultrasound image analysis could play a vital role in expanding the ultrasound's value by allowing healthcare providers with information to make clinical decisions, even without extensive experience interpreting these images [5].

Related Works: Recently, J. Born *et al.* [4] have open-sourced the POCUS dataset, which is the largest publicly available lung ultrasound dataset for COVID-19, and have leveraged a frame-based deep learning classifier to identify COVID-19 infection. Roy *et al.* [12] have developed a segmentation model to localize pathological artifacts in lung ultrasound images, and a video-based disease-severity predictor that scores individual frames, and then aggregates them with a soft approximation of the max function. Existing deep learning methods for detecting COVID-19 in lung ultrasound videos use a frame-based approach that loses valuable contextual and temporal information, which leaves significant room for improvement. Attention-based neural architectures have emerged as state-of-the-art models for video classification and captioning [9,10]. These video models typically need a lot of data to train, but large datasets of lung ultrasound videos for COVID-19 specifically do not yet exist. There is a need for a new approach that can incorporate knowledge across lung ultrasound datasets to achieve high performance on the COVID-19 classification task. While it is very important for clinicians to understand how deep learning models make their predictions, prior work in explainability for deep learning video-based models is limited. Existing methods for video model explainability are adapted from image model explainability methods, such as Grad-CAM [13] and Layer-Wise Relevance Propagation [3], which means that valuable information about the temporal features and motion are not captured in the explanations [8]. Model explainability is essential for the clinical application of deep learning methods, as it allows healthcare professionals to verify that the model is using sensible clinical factors to the make its prediction [2].

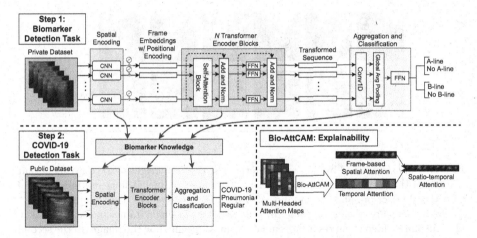

Fig. 1. Proposed framework: We propose an attention-based video model that is first trained on a pulmonary biomarker detection task, and an approach to transfer this knowledge to the COVID-19 prediction task. We also propose a novel method for generating attention-based video classification explanations called Bio-AttCAM to help clinicians understand which parts of the video the model is focusing on.

Contributions: In this paper, we propose an attention-based video classifier that leverages pulmonary biomarker knowledge transfer to achieve 80% precision and 87% recall on COVID-19, which outperforms the state-of-the-art model in COVID-19 diagnosis in lung ultrasound videos. Our pulmonary biomarker knowledge transfer approach guides the network to learn features that are highly relevant for COVID-19 diagnosis. Furthermore, our video model uses a transformer encoder to capture temporal information in the lung ultrasound videos. To our knowledge, this is the best achieved performance on this dataset. We also propose a novel method for generating video classification explanations called Biomarker Attention-scaled Class Activation Mapping (Bio-AttCAM). We demonstrate that our approach can assist clinicians with explanation maps that highlight key video features associated with the presence of COVID-19.

2 Materials and Methods

In this section, we will describe our lung ultrasound dataset and our proposed framework for the COVID-19 detection task. An overview of our proposed framework is shown in Fig. 1.

Dataset: In this work, we make use of two datasets. The first dataset is a private collection of 344 B-mode lung ultrasound cines from 69 patients across British Columbia, Canada, with ethics approval from the Institutional Medical Research Ethics Board. The cines were labelled by physicians for biomarkers that are commonly used for diagnosis of pulmonary disease, including COVID-19 pneumonia [6,11]. All cines were recorded with Clarius's convex ultrasound

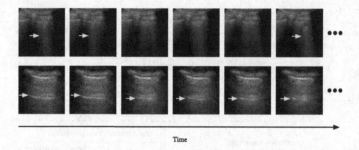

Time

Fig. 2. B-lines, specified by the images in the first row, are bright vertical artifacts that start from the pleural line and go downwards without fading. B-lines are indicative of COVID-19 infection, and the number of B-lines increases as the disease progresses [6]. These artifacts are often sparse and difficult to detect in single frames, but can be easily identified in a video given their movement. A-lines, specified by the images in the second row, are horizontal lines caused by the reverberation effect, where ultrasound waves travel back and forth between the transducer and the pleura. These are indicative of a healthy lung, as they are typically formed when the lung is not fluid-filled.

Table 1. Label distribution of the full-length cines in the private dataset and the public POCUS dataset.

Private Dataset	
A-lines	90
No A-lines	254
Total	344

Private Dataset	
B-lines	99
No B-lines	245
Total	344

Public POCUS Dataset	
COVID-19	68
Pneumonia	49
Regular	65
Total	182

transducers, as convex transducers are optimal to see deeply into the lungs of patients suspected of COVID-19 [6]. The data was collected using Clarius's HD transducers, including the C3HD, PA2HD, Vscan Extend, and CX50, which captured 165, 156, 22, and 1 cines respectively. These biomarkers include presence of A-lines, B-lines, lobar consolidations, and pleural irregularities. Our work focuses on A-lines and B-lines, as the present and absent cases of both artifacts appear in sufficient quantity in the dataset. Details about these artifacts can be viewed in Fig. 2. The second dataset is a publicly available dataset called POCUS consisting of B-mode lung ultrasound videos labelled as COVID-19, bacterial pneumonia, or regular [4]. The distribution of labels in both datasets can be viewed in Table 1. While A-lines correlate with healthy lungs and B-lines correlate with unhealthy lungs, COVID-19 diagnosis cannot be made solely based on the presence of these biomarkers.

Knowledge Transfer: In the process of pulmonary disease diagnosis, clinicians make their decisions based on the presence and severity of imaging biomarkers [6]. In previous works, COVID-19 detection was framed as a simple supervised classification problem. In that setting, where the amount of labeled data for COVID-19 is very limited, machine learning models are prone to overfitting, and often learn shortcuts instead of learning the real signatures of the disease. To overcome this problem, we leverage a different ultrasound dataset to guide the model to learn relevant features. This is done by first training the model to classify the presence of pulmonary imaging biomarkers from our private lung ultrasound dataset. Then, keeping these trained weights that incorporate pulmonary biomarker knowledge, the model's final layers are modified to match the output shape of the disease prediction problem. Lastly, all of the model's weights are fine-tuned on the POCUS dataset to detect COVID-19.

Attention-Based Video Model: We hypothesize that many individual lung ultrasound images are challenging for deep networks to label because of the sparsity of relevant imaging biomarkers in the videos, as well as the lack of motion information. Therefore, we propose an attention-based video model that uses a 2D convolutional neural network to capture per-frame spatial features, and then uses a transformer encoder to capture temporal information between frames. Next, we use a 1D convolutional layer over the temporal axis to capture additional temporal dependencies between neighboring features, followed by a 1D global average pooling layer to aggregate the sequence's features. Lastly, we add a fully connected layer before the softmax prediction heads. The attention mechanism allows the network to focus on the most relevant frames, which is particularly helpful for pulmonary disease diagnosis, as the sparsity of relevant imaging biomarkers in the video makes many frames unnecessary to focus on. The model architecture diagram can be found in Fig. 1.

Bio-AttCAM: Bio-AttCAM is a technique used to produce explanations of both the spatial and temporal factors that contributed to a model's decision. This is done by extracting the video model's transformer encoder attention weights to find the frames that the model pays more attention to. The temporal attention weights are used to scale each frame's Grad-CAM heatmap to show the relative importance of that frame. Lastly, we need to highlight the differences in attention and rescale the heatmap to be in the range $[0, 255]$. To do this, we divide by the maximum value in the heatmap, squash the values with a shifted sigmoid parametrized by the attention sparsity factor K, and then multiply by 255. The attention sparsity factor, K, modifies the explanation to accentuate important frames more, while suppressing the less important ones. Details about the algorithm can be found in Algorithm 1. Rather than simply looking at the model's activations to each frame's spatial features, our approach highlights the specific parts of the video with spatio-temporal features that most strongly contribute to the model's prediction. We believe that this approach works particularly well for lung disease diagnosis, as the imaging biomarkers for this task are often sparse and only show up on a few frames of the video. Due to the structure of the attention mechanism, the video model will be able to find these sparse frames, and Bio-AttCAM will visually highlight these features in the sparse behavior as desired.

Algorithm 1: Generate Bio-AttCAM heatmaps ϕ to visualize attention-based video model explanations.

Input: x, a video clip; f_θ, a CNN Transformer classification model with input video length T and h attention heads; K, the attention sparsity factor

1 $\phi \leftarrow$ empty list;
2 $g_\theta \leftarrow$ first layers of f_θ up to the first transformer encoder layer output, returns transformer's multi-headed attention weights;
3 $H \leftarrow g_\theta(x)$ list of h attention weight matrices of shape (T, T);
4 $H \leftarrow$ global_avg_pool_2d(H) across the h heads and the T attention weight matrix columns;
5 **for** $t = 1{:}T$ **do**
6      $G \leftarrow$ Grad-CAM(x_t, f_θ) visual explanation heatmap using spatial features;
7      $A \leftarrow H_t G$ Grad-CAM heatmap scaled by temporal attention weight;
8   Add A to list ϕ;
9 **end**
10 $\phi \leftarrow \phi$ / max(ϕ) rescale explanation heatmaps;
11 $\phi \leftarrow 255\sigma(K(\phi - 0.5))$ apply sigmoid with sparsity factor to ensure that the explanation accentuates important frames more, while suppressing the less important ones;
Output: ϕ

Table 2. Experiment implementation details.

Video Length	Frame Rate	Spatial Encoder	# Epochs	Batch Size	LR	Optimizer	K
5	9	VGG-16	40	8	1e-5	Adam	8

TF Version	Frame Size	# Transformer Blocks	GPU		Loss	
2.3.0	128x128	2	NVIDIA TITAN V		Weighted Categorical CE	

3 Experiments

Implementation Details: The experiment implementation details are summarized in Table 2. We evaluate our models using aggregated results from 5-fold cross validation and we compare our results to POCOVID-Net [4] as a baseline. We use their model architecture, but do not use the results from their paper, as their experimental results introduce optimization bias from their use of reduce learning rate and early stopping applied on their test sets. In contrast, we do not introduce optimization bias in this way.

Lung Biomarker Prediction Task: The first step of our proposed framework was to train a multi-headed network that would solve both the A-line and B-line prediction tasks. These tasks are both binary classification problems, with the goal of classifying the biomarker as present or absent. This model could then be transferred to the COVID-19 prediction task. We trained and compared two different deep network architectures for this task. The first architecture is our baseline model called POCOVID-Net, a frame-based averaging classification model that was proposed by [4]. We use the same POCOVID-Net model architecture, but change the last two layers to get the desired output shape for this multi-headed classification task. The second architecture is our proposed attention-based video classification model that uses the same spatial encoder as POCOVID-Net, but uses a transformer encoder to capture temporal information across frames, rather than simply averaging the predictions of all frames. The results for this experiment can be viewed in Table 3, where our proposed multi-headed CNN transformer model shows the best performance.

Table 3. Comparison of POCOVID-Net and CNN Transformer for the imaging biomarker detection multi-task problem. The precision, recall, and F1-score values are macro averages between the present class (presence of imaging biomarker) and the absent class (absence of imaging biomarker). The CNN transformer achieves better performance than POCOVID-Net for both tasks, as it is able to leverage temporal information and focus on specific frames of interest.

	Task	Precision	Recall	F1-score
POCOVID-Net [4]	A-lines	85%	83%	84%
	B-lines	71%	76%	73%
CNN Transformer (ours)	A-lines	**86%**	**84%**	**85%**
	B-lines	**78%**	**80%**	**79%**

Knowledge Transfer to COVID-19 Prediction: The next step of our experiment is to transfer the imaging biomarker knowledge from the models in the previous section to the COVID-19 prediction task. We start with the same model architecture and weights from the previous section to guide the network to find the spatio-temporal features relevant for COVID-19. Next, we change the last two layers to allow the model to find the combination of these features that are indicative of COVID-19 infection and then fine-tune all of the weights on the POCUS dataset. We perform an ablation study to demonstrate the effect of the attention-based video model and the biomarker knowledge transfer to COVID-19 diagnosis on the POCUS dataset. We first evaluate the baseline POCOVID-Net model and our proposed CNN transformer model without knowledge transfer. Then, we use the models from the previous section to evaluate POCOVID-Net and the CNN transformer with knowledge transfer. The results for this experiment can be viewed in Table 4, where our proposed CNN transformer model with knowledge transfer outperform the other models, and the improvement was statistically significant.

Bio-AttCAM Qualitative: We perform a qualitative analysis comparing Bio-AttCAM with Grad-CAM. Comparing the heatmaps shown in Fig. 3, we demonstrate that Bio-AttCAM can generate heatmaps that highlight the key frames of a video and suppress frames that the model does not find important. This is particularly relevant for lung ultrasound videos, as they are typically lower quality and noisier than other imaging modalities. This would allow clinicians to more quickly understand and evaluate if the model was focusing on clinically meaningful factors. Using Grad-CAM, each frame's heatmap is treated equally, so it fails to highlight the key frames more than the others.

Table 4. Ablation study demonstrating the effect of the CNN transformer and the imaging biomarker knowledge transfer over the baseline POCOVID-Net model. Both contributions show an improvement to the performance, but the CNN transformer addition makes a larger improvement.

	Class	Precision	Recall	F1-score
POCOVID-Net [4]	COVID-19	67%	76%	71%
	Pneumonia	49%	83%	61%
	Regular	92%	65%	76%
CNN transformer	COVID-19	**86%**	82%	**84%**
	Pneumonia	62%	**95%**	75%
	Regular	**95%**	82%	88%
POCOVID-Net w/ knowledge	COVID-19	74%	78%	76%
	Pneumonia	58%	78%	67%
	Regular	**95%**	81%	88%
CNN transformer w/ knowledge (ours)	COVID-19	80%	**87%**	83%
	Pneumonia	**86%**	93%	**90%**
	Regular	93%	**85%**	**89%**

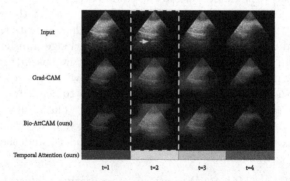

Fig. 3. Four frames of a lung ultrasound video containing sparse A-lines that are most prominent at $t = 2$, and the explanation heatmaps generated by Grad-CAM and Bio-AttCAM. The temporal attention values refers to the relative attention to put on each frame. Bio-AttCAM successfully highlights the key frame at $t = 2$, as this frame has the largest temporal attention value, which is consistent with independent key frame selections from two clinicians. Grad-CAM treats all frames equally, so it fails to highlight the key frame more than the others.

Bio-AttCAM Quantitative: Two clinicians were asked to select the top three most important and least important frames for their labeling across 16 videos. We compare the temporal attention weights from Bio-AttCAM to their frame selections. Our temporal explanations were able to capture at least one of the key frames in $\frac{13}{16}$ videos and not capture any of the uninformative frames in $\frac{15}{16}$, which demonstrates the efficacy of this method to highlight the key frames and suppress the uninformative frames in videos.

4   Discussion and Conclusion

In this paper, we outperformed the state-of-the-art model for COVID-19 lung ultrasound diagnosis through imaging biomarker knowledge transfer with a CNN transformer model, and proposed a novel method for video explainability called Bio-AttCAM. Our model achieves 80% precision and 87% recall for COVID-19 (13% and 11% improvement to the state-of-the-art respectively). From our ablation study, we found that the individual additions of transformer encoders and biomarker pretraining result in an average increase of 13% and 7.67% increase to F1-score across classes, respectively (all individual F1-scores improved). These results suggest that the transformer encoders allow the model to capture additional temporal features to better detect COVID-19, and that biomarker pretraining improves the model's performance by guiding it to learn clinically relevant features. We believe biomarker pretraining is helpful for small datasets, where the model is prone to finding shortcuts and learning irrelevant features. The attention mechanism allows the model to focus on the key frames that are important for diagnosis and then generate explanation heatmaps with Bio-AttCAM, which is useful for videos whose important features are sparse between frames. Future work will be focused on extending the number of biomarkers incorporated for the knowledge transfer with the availability of more data.

Acknowledgement. This work was supported by Canada's Digital Technology Supercluster, the Natural Sciences and Engineering Research Council of Canada (NSERC), and the Canadian Institutes of Health Research (CIHR).

References

1. Abrams, E.R., Rose, G., Fields, J.M., Esener, D.: Point-of-care ultrasound in the evaluation of COVID-19. J. Emerg. Med. **2020**, 403–408 (2020)
2. Amann, J., Blasimme, A., Vayena, E., Frey, D., Madai, V.: Explainability for artificial intelligence in healthcare: a multidisciplinary perspective. BMC Med. Inform. Decision Making **20**, 310 (2020)
3. Bach, S., et al.: On pixel-wise explanations for non-linear classifier decisions by layer-wise relevance propagation. PLoS ONE **10**, 1–46 (2015)
4. Born, J., et al.: Accelerating detection of lung pathologies with explainable ultrasound image analysis. Appl. Sci. **11**(2), 672 (2021)
5. Brattain, L.J., Telfer, B.A., Dhyani, M., Grajo, J.R., Samir, A.E.: Machine learning for medical ultrasound: status, methods, and future opportunities. Abdom Radiol. (NY) **2018**, 786–799 (2018)
6. Buda, N., Segura-Grau, E., Cylwik, J., Wełnickid, M.: Lung ultrasound in the diagnosis of COVID-19 infection - a case series and review of the literature. Adv. Med. Sci. **2020**, 378–385 (2020)
7. Dong, E., Du, H., Gardner, L.: An interactive web-based dashboard to track Covid-19 in real time. Lancet Infect. Dis. **20**(5), 533–534 (2020)

8. Hiley, L., Preece, A., Hicks, Y.: Explainable deep learning for video recognition tasks: a framework & recommendations. arXiv preprint arXiv:1909.05667 (2019)

9. Li, J., Qiu, H.: Comparing attention-based neural architectures for video captioning (2019)

10. Liu, T., Shao, Q.: BERT for large-scale video segment classification with test-time augmentation. arXiv preprint arXiv:1912.01127 (2019)

11. Peng, Q., et al.: Findings of lung ultrasonography of novel corona virus pneumonia during the 2019–2020 epidemic. Intensive Care Med. **46**, 849–850 (2020)

12. Roy, S., et al.: Deep learning for classification and localization of COVID-19 markers in point-of-care lung ultrasound. IEEE Trans. Med. Imag. **39**(8), 2676–2687 (2020)

13. Selvaraju, R.R., et al.: Grad-CAM: visual explanations from deep networks via gradient-based localization. In: 2017 IEEE International Conference on Computer Vision (ICCV), pp. 618–626 (2017)

Endoscopic Ultrasound Image Synthesis Using a Cycle-Consistent Adversarial Network

Alexander Grimwood[1,2], Joao Ramalhinho[1,2], Zachary M. C. Baum[1,2],
Nina Montaña-Brown[1,2], Gavin J. Johnson[3], Yipeng Hu[1,2], Matthew J. Clarkson[1,2],
Stephen P. Pereira[4], Dean C. Barratt[1,2], and Ester Bonmati[1,2(✉)]

[1] Wellcome / EPSRC Centre for Interventional and Surgical Sciences,
University College London, London, UK
e.bonmati@ucl.ac.uk
[2] UCL Centre for Medical Image Computing, University College London, London, UK
[3] Department of Gastroenterology, University College London Hospital, London, UK
[4] Institute for Liver and Digestive Health, University College London, London, UK

Abstract. Endoscopic ultrasound (EUS) is a challenging procedure that requires skill, both in endoscopy and ultrasound image interpretation. Classification of key anatomical landmarks visible on EUS images can assist the gastroenterologist during navigation. Current applications of deep learning have shown the ability to automatically classify ultrasound images with high accuracy. However, these techniques require a large amount of labelled data which is time consuming to obtain, and in the case of EUS, is also a difficult task to perform retrospectively due to the lack of 3D context. In this paper, we propose the use of an image-to-image translation method to create synthetic EUS (sEUS) images from CT data, that can be used as a data augmentation strategy when EUS data is scarce. We train a cycle-consistent adversarial network with unpaired EUS images and CT slices extracted in a manner such that they mimic plausible EUS views, to generate sEUS images from the pancreas, aorta and liver. We quantitatively evaluate the use of sEUS images in a classification sub-task and assess the Fréchet Inception Distance. We show that synthetic data, obtained from CT data, imposes only a minor classification accuracy penalty and may help generalization to new unseen patients. The code and a dataset containing generated sEUS images are available at: https://ebonmati.github.io.

Keywords: Endoscopic ultrasound · Synthesis · Classification

1 Introduction

Endoscopic Ultrasound (EUS) is a minimally-invasive procedure to assess the gastrointestinal tract including pancreatobiliary disorders such as pancreatic cancer. It is a complex procedure that combines ultrasound and endoscopy, requiring advanced cognitive and technical skills, such as ultrasound image interpretation [1].

Recent advances in machine learning have made it feasible to automatically classify images and identify standard planes, which can improve ultrasound (US) image

© Springer Nature Switzerland AG 2021
J. A. Noble et al. (Eds.): ASMUS 2021, LNCS 12967, pp. 169–178, 2021.
https://doi.org/10.1007/978-3-030-87583-1_17

interpretation and assist clinicians during navigation with the aim to make the diagnosis more accurate. However, the success of deep-learning based applications relies on the acquisition of a large, well-curated dataset with enough quality to be representative and useful. This is a big challenge in US applications as often training data is limited and models tend to have overfitting problems [2]. Data acquisition during EUS procedures is especially difficult and demanding due to the disruption caused and time required by real-time labelling, as well as the inaccuracies associated with retrospective labelling, because it is difficult to confidently identify landmarks without the 3D spatial and temporal context.

Medical image synthesis using convolutional neural networks (CNN) has been shown to be able to successfully translate Magnetic Resonance Imaging (MRI) to Computed Tomography (CT) [3] and to translate US to MRI [4]. In this work, we evaluate the use of a cycle-consistent adversarial network (CycleGAN) [5] to perform CT-to-EUS image translation to generate synthetic EUS (sEUS) images for the purpose of data augmentation. As an example, CycleGANs have been used before to improve the realism in US simulation from CT in a ray-casting approach, or to generate labelled US images from musculoskeletal US as a data augmentation strategy [6, 7]. The CycleGAN approach is of particular interest for our clinical application as no commercially available endoscopes exist capable of acquiring paired US/CT data, making endoscopy training and patient navigation difficult. The aim of our study is: 1) to assess the similarity between real (EUS) and synthetic (sEUS) images, and 2) to evaluate the use of sEUS images as a data augmentation strategy in a clinically relevant EUS classification task.

2 Methods

2.1 Data

CT Data. CT data from five patients was obtained, four were from the MICCAI 2015 workshop and challenge: Multi-Altas Labelling Beyond the Cranial Vault [8]. CT slice dimensions were 512×512 with pixel sizes from 0.59 mm to 0.73 mm. Slice thicknesses were 3 mm, with volume depths from 393 mm to 444 mm. We also included a CT volume of size $512 \times 512 \times 229$ with pixel dimensions of 0.55×0.55 mm and a slice thickness of 1 mm. Segmentations of the following structures were available for this study: stomach, pancreas, liver and aorta.

EUS Data. EUS images were obtained from five patients who underwent an EUS-guided examination at University College Hospital London. Data were acquired from a Hitachi Preirus EUS console and a Pentax EG-3270UK or EG-3870UTK US linear video endoscopes with a 7.5 MHz probe. EUS images were collected from video frames of each examination recorded with a resolution of 720×480 pixels at imaging depths from 4 mm to 6 mm and cropped to 522×200 pixels, removing identifiable text and depth-attenuated regions. Anatomical landmarks were identified by an expert and recorded during the procedure. EUS images containing the three clinically relevant anatomical landmarks: pancreas, liver and aorta, were manually identified and collected. Assorted images outside of these labels were also collected for a background class used in the classification subtask. Images from four patients were used for CycleGAN and classifier training. Data from the remaining patient were used to evaluate classification performance. Table 1 shows a summary of the number of images and available labels from each patient.

Table 1. Summary of EUS images collected for training the CycleGAN and classifier, and for evaluation of the classifier.

Patient	Task	Images	Labels			
			Aorta	Liver	Pancreas	Background
EUS1	train	3299	0	0	1694	1605
EUS2	train	4758	767	0	3991	0
EUS3	eval	4451	1301	729	2421	0
EUS4	train	4996	1126	1485	2385	0
EUS5	test	1933	141	503	1289	0

2.2 EUS/CT Image-To-Image Translation

Synthetic EUS images were generated using CycleGANs trained to translate 2D CT image planes into sEUS images (Fig. 1). The CT plane locations, orientations and bounding dimensions approximated real EUS views. Candidate sEUS locations were automatically identified in CT volumes using the associated CT segmentation labels. Points were randomly sampled along the outer surface of the CT stomach segmentation. Realistic sEUS probe orientations were identified at each point by randomly generating poses within a 30° cone normal to the stomach and retaining only poses where the view intersected an anatomical label of interest (i.e., aorta, liver or pancreas) as shown in Fig. 2. These poses were recorded as transformation matrices and saved to file.

During CycleGAN training, 2D CT images were sliced from CT volumes on the fly using a previously reported simulation pipeline [9]. The framework extracted a sEUS field of view, defined by a transformation matrix, in the CT volume. The CT planes were then passed to the CycleGAN with randomly selected EUS images as an unpaired input dataset.

The CycleGAN was based upon a previously described implementation comprising a generator and adversarial discriminator for each imaging modality [5]. Paired EUS and CT plane images were passed to their respective generators, which were trained to map their input modality into synthetic images (i.e., CT to sEUS and EUS to sCT). These synthetic images were subsequently passed to the relevant generator for mapping back to their original modalities. Training was governed by adversarial losses calculated at each discriminator and by cycle consistency losses comparing input images to those mapped to a synthetic modality and then remapped back to their original.

A small pre-trained Gaussian denoising network was added before each discriminator to prevent the generator from embedding information capable of facilitating loss minimization without improving image-to-image translation [10, 11].

2.3 Implementation Details

Three CycleGANs were trained, one for each of the three labels: aorta, liver, and pancreas. A batch size of 1 was used for 200 epochs, where the number of iterations per epoch was limited by the relevant EUS dataset size. Other hyperparameters were set to the defaults used in the original implementation [5]. Adam optimization was used on discrimination and generation networks with a learning rate of 0.0002 that decayed linearly after 100 epochs [12]. Image intensity values were normalized between −1 and 1. Data augmentation was applied to all images, incorporating random horizontal flips and random cropping within a 40 pixel margin. The CT plane slicer and CycleGAN frameworks were run simultaneously on a single 16 GB NVIDIA Quadro P5000 GPU. CT slices were selectively generated so that EUS images were paired only with CT slices containing >1000 labelled pixels and <50 pixels with high Hounsfield Units, indicative of bone. For each epoch, a 90/10 training/validation split was randomly applied to the dataset. Losses were plotted against epoch and inspected to ensure convergence was achieved. All models were implemented in TensorFlow 2.2 and CUDA Toolkit 10.1 [13]. Synthetic data for the classifier evaluation sub-task was created using the trained CT-sEUS generator network from each CycleGAN. Open-source code was used where possible and is available at: https://ebonmati.github.io.

2.4 Evaluation

Evaluating GANs remains an open challenge, as there is no concrete way to quantify how realistic and diverse the synthetic images are, and no ground truth exists. Often, models are evaluated in a subjective and quantitative manner by asking several observers to rate the images [14]. In this work, we used the Fréchet Inception Distance and a classification sub-task to evaluate our model, as described below.

Fréchet Inception Distance. To quantitatively evaluate the quality of the synthetic images, we calculated the Fréchet Inception Distance (FID) [15]. FID is a widely used metric for evaluating the similarity between the generated images (synthetic) and the real images. FID uses the activation distributions of the Inception-v3 model [16] to calculate the distance between real and synthetic images. We used the pre-trained Inception-v3 model available in Keras [16] to obtain the activation distributions for our real and synthetic images, where the FID score was then calculated as follows:

$$FID = \|\mu_X - \mu_Y\|^2 + Tr\left(\Sigma_X + \Sigma_Y - 2\sqrt{\Sigma_X \Sigma_Y}\right), \tag{1}$$

where μ_X and μ_Y are the mean of the feature vectors for the real and synthetic images, respectively; Σ_X and Σ_Y are the covariance matrix for the real and synthetic images, respectively; $\|\mu_X - \mu_Y\|^2$ refers to the sum squared difference between the two mean vectors, and Tr is the trace. A lower FID indicates better-quality synthetic images; conversely, a higher score indicates a lower-quality image. An FID of 0 demonstrates that the activation distribution of the synthetic images is identical to that of the real images. FID is also capable of detecting intra-class mode dropping (i.e., a model that generates only one type of image for each landmark or class), noise, blurring, and other systematic distortions.

Classification Sub-task. We evaluate the use of the synthetic EUS images in a classification sub-task. The aim here is to: 1) evaluate the use of synthetic EUS images to classify real EUS images, and 2) to use the synthetic EUS images as a data augmentation strategy. As summarized in Table 2, the number of real training images for each class was: 3,194 aorta, 2,214 liver and 10,491 pancreas. A fourth background class was added in training only, incorporating 1,605 EUS images from a mix of indiscernible anatomy and poor quality images. To achieve this, we implemented a simple VGG-16 classification model to classify EUS into the following classes: aorta, liver and pancreas. We used the pre-trained weights from ImageNet, a batch size of 64, a learning rate of $1e^{-7}$ and 100 epochs. As loss function, we used a weighted categorical cross entropy with the weights of 4.23, 4.40, 4.48, and 2.78 for aorta, background, liver and pancreas, respectively. We trained the model using 5 different ratios of synthetic/real images: 0% synthetic +100% real, 25% synthetic +75% real, 50% synthetic +50% real, 75% synthetic and 25% real. For each synthetic ratio, we report the accuracy, precision, recall and F1-measure. Pairs of classifier models were compared using McNemar tests to assess whether differences in accuracy were significant.

Qualitative Evaluation. We are also interested in the visual explanation and spatial localization of important regions in the EUS and sEUS images that were used to predict the corresponding class. We used the Gradient-weighted Class Activation Mapping (Grad-CAM) to generate the class activation maps for each sample [17]. These maps provide an insight into the model interpretation by backpropagating the gradients from the last convolutional layer.

3   Results and Discussion

Figure 1 shows a comparison between a real EUS image and a sEUS image for each of the anatomical landmarks selected (aorta, liver, pancreas). Visually inspecting the generated sEUS images, we observed that the sEUS images obtained with CycleGAN look realistic as the main features of the anatomical landmarks are preserved.

In Table 2 we report the classifier performance when trained on varying ratios of sEUS to real EUS images, with the number of sEUS increasing with the ratio. From this table we can observe that classification accuracy is maintained for sEUS ratios up to 75%. We attribute this to the fact sEUS images may provide a consistent representation of patient variation on the selected anatomical landmarks, making it feasible to generalize to new patients. Liver F-measures were consistently low, indicating poor classification performance and degrading overall accuracy scores. Due to the liver's size and position, liver-labelled images can often contain additional anatomical features belonging to the other classes. We speculate this may be a contributing factor to the consistently low F-measures.

Single-sided McNemar tests comparing classifier pairs indicated the small reduction in accuracy, from 0.63 to 0.61 with increasing sEUS ratio, was statistically significant ($p < 0.05$).

The FID scores are shown in Table 3. Although the FID measure is widely used to evaluate the realism of synthetic images, a significant limitation arises from its reliance on a pre-trained model (ImageNet) that does not comprehensively represent US-specific features. This lack of accurate representation is compounded by the small dataset used in this study (and in medical imaging generally) compared to the originally intended application of FID. As such, we cannot expect the predicted activation distributions to provide authoritative results on our specific clinical application. An indication of the ideal FID is given by the differences between random subsets within the EUS data (as shown in the EUS vs EUS results). To have a better quantification of a bad FID value, we compared all the EUS images to all the EUS images with added noise using a Gaussian distribution with 0 mean and a standard deviation of 0.1. Our synthetic images achieved lower scores in comparison to that from noisy US images which yielded a FID > 300.

Finally, Fig. 3 shows the Grad-CAM activations for two image examples (one EUS and one sEUS) of the pancreas. Note the model has focused on the area representing the pancreas to make a correct prediction.

Other studies have used GAN-based methods to simulate and augment ultrasound image data: Bargsten and Schlaefer developed SpeckleGAN, which generates intra-venous ultrasound speckle simulations from segmentation maps, achieving FID scores <115 [18]. Peng, et al. generated synthetic ultrasound from MRI images and qualitatively demonstrated their equivalence to numerical simulations [19]. A broader examination of GAN-based approaches in medical imaging was presented by Yi et al. [20].

In future, this study could be extended to aid EUS navigation by establishing a real-time image labelling and automated landmark recognition framework, for example, by using the Grad-CAM maps to localise salient features in EUS video, as demonstrated in this work. Further potential enhancements include developing a single CycleGAN model capable of generating all three landmark types to enable multi-class object generation and detection.

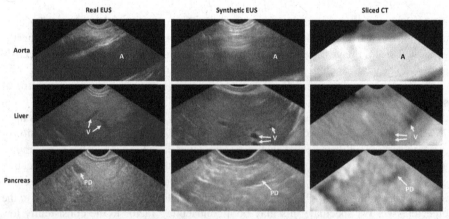

Fig. 1. Comparison between a real EUS image, a synthetic EUS image and the sliced CT from which it was generated for each anatomical landmark (aorta, liver and pancreas). Indicative anatomical features are shown in red: A – aorta, V – liver vasculature, PD – pancreatic duct. (Color figure online)

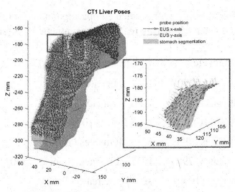

Fig. 2. Subset of candidate EUS probe positions and orientations at the stomach surface for liver views within a CT volume.

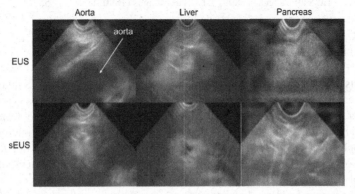

Fig. 3. Normalized class activation maps for real EUS images and a synthetic EUS images (sEUS) representing the aorta, liver and pancreas. The red areas represent increased regions of activation used by the model to make a correct prediction. (Color figure online)

Table 2. Classifier performance when trained on varying ratios of synthetic to real EUS images. The number of sEUS images increases with synthetic ratio.

Synthetic ratio (%)	Precision	Recall	F1-measure			Accuracy
			Aorta	Liver	Pancreas	
0	0.46	0.51	0.38	0.07	0.77	0.63
25	0.45	0.54	0.38	0.05	0.78	0.62
50	0.44	0.54	0.38	0.05	0.78	0.62
75	0.43	0.53	0.36	0.06	0.77	0.61

Table 3. Fréchet inception distance scores when comparing random subsets of real EUS images within the same class, and when comparing real EUS to synthetic EUS (sEUS) images.

Control (n images)	Compared to (n images)	FID
EUS pancreas (5250)	EUS pancreas (5241)	2.00
EUS aorta (1599)	EUS aorta (1595)	6.96
EUS liver (1110)	EUS liver (1104)	11.03
EUS all images (7959)	EUS all images (7940)	1.83
EUS all images (7959)	EUS all images + noise (7940)	312.56
EUS pancreas (10491)	sEUS pancreas (11763)	79.88
EUS aorta (3194)	sEUS aorta (2774)	71.30
EUS liver (2214)	sEUS liver (4365)	71.68
EUS all images (15899)	sEUS all images (18902)	55.31

4 Conclusions

The results of this work demonstrate that the generation of synthetic EUS images, from CT data, can support training of a simple classification model when data is scarce as it may better represent the population. It allows generation of a large dataset from specific anatomical landmarks that are relevant for the clinical application of interest, which would not be possible otherwise (as demonstrated by the poor accuracy obtained when using the only real EUS data available). The proposed method is easy to use compared to manual data acquisition and labelling, which is a task that is time consuming and requires the input of a clinical expert.

Acknowledgements. This work is supported by the Wellcome/EPSRC Centre for Interventional and Surgical Sciences (WEISS) (203145/Z/16/Z) and by Cancer Research UK (CRUK) Multi-disciplinary Award (C28070/A19985). NMB is supported by the EPSRC-funded UCL Centre for Doctoral Training in Intelligent, Integrated Imaging in Healthcare (i4health) (EP/S021930/1). ZMC Baum is supported by the Natural Sciences and Engineering Research Council of Canada Postgraduate Scholarships-Doctoral Program, and the UCL Overseas and Graduate Research Scholarships. SP Pereira was supported by the UCLH/UCL Comprehensive Biomedical Centre, which receives a proportion of funding from the Department of Health's National Institute for Health Research (NIHR) Biomedical Research Centres funding scheme.

References

1. Bonmati, E., et al.: Determination of optimal ultrasound planes for the initialisation of image registration during endoscopic ultrasound-guided procedures. Int. J. Comput. Assist. Radiol. Surg. **13**(6), 875–883 (2018). https://doi.org/10.1007/s11548-018-1762-2
2. Liu, S., et al.: Deep learning in medical ultrasound analysis: a review (2019). https://doi.org/10.1016/j.eng.2018.11.020

3. Nie, D., et al.: Medical image synthesis with context-aware generative adversarial networks. In: Descoteaux, M., Maier-Hein, L., Franz, A., Jannin, P., Collins, D.L., Duchesne, S. (eds.) MICCAI 2017. LNCS, vol. 10435, pp. 417–425. Springer, Cham (2017). https://doi.org/10.1007/978-3-319-66179-7_48

4. Jiao, J., Namburete, A.I.L., Papageorghiou, A.T., Noble, J.A.: Self-supervised ultrasound to MRI fetal brain image synthesis. IEEE Trans. Med. Imaging. **39**, 4413–4424 (2020). https://doi.org/10.1109/TMI.2020.3018560

5. Zhu, J.-Y., Park, T., Isola, P., Efros, A.A.: Unpaired image-to-image translation using cycle-consistent adversarial networks. In: Proceedings IEEE International Conference Computer Vision, October 2017, pp. 2242–2251 (2017)

6. Zhang, L., Portenier, T., Goksel, O.: Learning ultrasound rendering from cross-sectional model slices for simulated training. Int. J. Comput. Assist. Radiol. Surg. **16**(5), 721–730 (2021). https://doi.org/10.1007/s11548-021-02349-6

7. Cronin, N.J., Finni, T., Seynnes, O.: Using deep learning to generate synthetic B-mode musculoskeletal ultrasound images. Comput. Methods Programs Biomed. **196**, 105583 (2020). https://doi.org/10.1016/j.cmpb.2020.105583

8. Landman, B., Xu, Z., Igelsias, J.E., Styner, M., Langerak, T.R., Klein, A.: Multi-atlas labeling beyond the cranial vault. https://doi.org/10.7303/syn3193805

9. Ramalhinho, J., Tregidgo, H.F.J., Gurusamy, K., Hawkes, D.J., Davidson, B., Clarkson, M.J.: Registration of untracked 2D laparoscopic ultrasound to CT images of the liver using multi-labelled content-based image retrieval. IEEE Trans. Med. Imaging. **40**, 1042–1054 (2021). https://doi.org/10.1109/TMI.2020.3045348

10. Porav, H., Musat, V., Newman, P.: Reducing steganography in cycle-consistency GANs. In: Proceedings of the IEEE/CVF Conference on Computer Vision and Pattern Recogni-tion (CVPR) Workshops, pp. 78–82 (2019)

11. Zhang, K., Zuo, W., Chen, Y., Meng, D., Zhang, L.: Beyond a Gaussian Denoiser: residual learning of deep CNN for Image Denoising. IEEE Trans. Image Process. **26**, 3142–3155 (2017). https://doi.org/10.1109/TIP.2017.2662206

12. Kingma, D.P., Ba, J.: Adam: a method for stochastic optimization. arXiv preprint arXiv:1412.6980. 22 Dec 2014. https://arxiv.org/abs/1412.6980

13. Abadi, M., et al.: TensorFlow: large-scale machine learning on heterogeneous systems (2015). https://www.tensorflow.org/ https://doi.org/10.5281/zenodo.4724125

14. Lucic, M., Kurach, K., Michalski, M., Gelly, S., Bousquet, O.: Are GANs Created Equal? A large-scale study. In: Bengio, S., Wallach, H., Larochelle, H., Grauman, K., Cesa-Bianchi, N., Garnett, R. (eds.) Advances in Neural Information Processing Systems. Curran Associates, Inc. (2018)

15. Heusel, M., Ramsauer, H., Unterthiner, T., Nessler, B., Hochreiter, S.: GANs trained by a two time-scale update rule converge to a local Nash equilibrium. In: Proceedings of the 31st International Conference on Neural Information Processing Systems, pp. 6629–6640 (2017). https://doi.org/10.5555/3295222.3295408

16. Szegedy, C., Vanhoucke, V., Ioffe, S., Shlens, J., Wojna, Z.: Rethinking the inception architecture for computer vision. In: Proceedings of the IEEE Computer Society Conference on Computer Vision and Pattern Recognition, pp. 2818–2826. IEEE Computer Society (2016). https://doi.org/10.1109/CVPR.2016.308

17. Selvaraju, R.R., Cogswell, M., Das, A., Vedantam, R., Parikh, D., Batra, D.: Grad-CAM: visual explanations from deep networks via gradient-based localization. Int. J. Comput. Vision **128**(2), 336–359 (2019). https://doi.org/10.1007/s11263-019-01228-7

18. Bargsten, L., Schlaefer, A.: SpeckleGAN: a generative adversarial network with an adaptive speckle layer to augment limited training data for ultrasound image processing. Int. J. Comput. Assist. Radiol. Surg. **15**(9), 1427–1436 (2020). https://doi.org/10.1007/s11548-020-02203-1

19. Peng, B., Huang, X., Wang, S., Jiang, J.: A real-time medical ultrasound simulator based on a generative adversarial network model. In: 2019 IEEE International Conference on Image Processing (ICIP), pp. 4629–4633 (2019). https://doi.org/10.1109/ICIP.2019.8803570
20. Yi, X., Walia, E., Babyn, P.: Generative adversarial network in medical imaging: a review. Med. Image Anal. **58**, 101552 (2019). https://doi.org/10.1016/j.media.2019.101552

Realistic Ultrasound Image Synthesis for Improved Classification of Liver Disease

Hui Che[1], Sumana Ramanathan[1], David J. Foran[4], John L. Nosher[2], Vishal M. Patel[3], and Ilker Hacihaliloglu[1,2(✉)] 🆔

[1] Department of Biomedical Engineering, Rutgers University, Piscataway, NJ, USA
ilker.hac@soe.rutgers.edu
[2] Department of Radiology, Rutgers Robert Wood Johnson Medical School, New Brunswick, NJ, USA
[3] Department of Electrical and Computer Engineering, Johns Hopkins University, Baltimore, MD, USA
[4] Rutgers Cancer Institute of New Jersey, New Brunswick, NJ, USA

Abstract. With the success of deep learning-based methods applied in medical image analysis, convolutional neural networks (CNNs) have been investigated for classifying liver disease from ultrasound (US) data. However, the scarcity of available large-scale labeled US data has hindered the success of CNNs for classifying liver disease from US data. In this work, we propose a novel generative adversarial network (GAN) architecture for realistic diseased and healthy liver US image synthesis. We adopt the concept of stacking to synthesize realistic liver US data. Quantitative and qualitative evaluation is performed on 550 in-vivo B-mode liver US images collected from 55 subjects. We also show that the synthesized images, together with real in vivo data, can be used to significantly improve the performance of traditional CNN architectures for Nonalcoholic fatty liver disease (NAFLD) classification.

Keywords: Nonalcoholic fatty liver disease · Ultrasound · Classification · Stacked generative adversarial network · Deep learning

1 Introduction

Nonalcoholic fatty liver disease (NAFLD) is being recognized as one of the most common liver diseases worldwide, affecting up to 30% of the adult population in the Western countries [25]. It is defined as a condition with increased fat deposition in the hepatic cells due to obesity and diabetes in the absence of alcohol consumption [4]. Patients with NAFLD are at an increased risk for the development of cirrhosis and hepatocellular carcinoma (HCC) which is one of the fastest-growing causes of death in the United States and poses a significant economic burden on healthcare [19]. Therefore, early diagnosis of NAFLD is important for improved management and prevention of HCC. Liver biopsy is considered the gold standard for diagnosing NAFLD [8]. However, biopsy is an

© Springer Nature Switzerland AG 2021
J. A. Noble et al. (Eds.): ASMUS 2021, LNCS 12967, pp. 179–188, 2021.
https://doi.org/10.1007/978-3-030-87583-1_18

invasive and expensive procedure associated with serious complications making it impractical as a diagnostic tool [24]. Incorrect staging in 20% of the patients has also been reported due to sampling error and/or inter-observer variability [24]. Diagnostic imaging, based on Ultrasound (US), Magnetic resonance imaging (MRI), and Computed Tomography (CT), has been utilized as a safe alternative. Due to being cost-effective, safe, and able to provide real-time bedside imaging US has been preferred over MRI and CT [14]. Nonetheless, studies have shown that the specificity and sensitivity of US to detect the presence of steatosis is very poor [12]. Furthermore, the appearance of the tissue can be very easily affected by machine acquisition settings and the experience of the clinicians [1,22].

To overcome the drawbacks of US imaging and improve clinical management of liver disease, Computer-Aided Diagnostic (CAD) systems have been developed. With the success of deep learning methods in the analysis of medical images, recent focus has been on the incorporation of convolutional neural networks (CNNs) into CAD systems to improve the sensitivity and specificity of US in diagnosing liver disease [5–7,15,16,21]. Although successful results were reported, one of the biggest obstacles hindering the improved adaptation of deep learning-based CAD systems in clinical practice is the unavailability of large-scale annotated datasets. The collection of large-scale annotated medical data is a very expensive and long process. Albeit there are many publicly available datasets and world challenges, the data available is still limited and the focus has been on certain clinical applications based on MRI, CT, and X-ray imaging. One of the most commonly used practices to overcome the scarce data problem is data augmentation based on image geometric transformation techniques such as rotation, translation, and intensity transformations [2,7,17]. However, these transformations result in images with similar feature distributions and do not increase the diversity of the dataset required to improve the performance of any CNN model. A new type of data augmentation is image synthesis using generative adversarial networks (GANs). GAN-based medical image synthesis has become popular for improving the dataset size and has been extensively investigated for improving classification and segmentation tasks [11,27]. However, to the best of our knowledge GAN-based image synthesis in the context of liver disease classification from US data has not been investigated previously.

In this work, a novel GAN-based deep learning method is proposed to synthesize B-mode liver US data. Our contributions include: 1- We propose a stacked GAN architecture for realistic liver US image synthesis. 2- Using ablation studies we show how the performance of state-of-the-art GAN architectures can be improved using the proposed stacked GAN architecture. Qualitative and quantitative evaluations are performed on 550 B-mode in-vivo liver US images collected from 55 subjects. Extensive experiments demonstrate our method improves traditional single-stage GANs to generate B-mode liver US data. We also show that by using a larger balanced NAFLD dataset, including real and synthesized data, the performance of the liver disease classification task can be improved by 4.34%.

2   Materials and Methods

The architecture of our proposed network is shown in Fig. 1. Specifically we design a stacked GAN model (StackGAN) to generate high-resolution liver US images in two stages. A common GAN layout is utilized in Stage-I to synthesize a mid-resolution image. Stage-II GAN, the main contribution of this work, aims to output an image with improved tissue details by integrating features from the mid-resolution image during the generative process. The two stage approach overcomes the training instability of single stage GANs and results in improved and realistic representation of the synthesized liver US data. The overall network produces high-resolution images using random noise as input, not relying on any prior information from the original data.

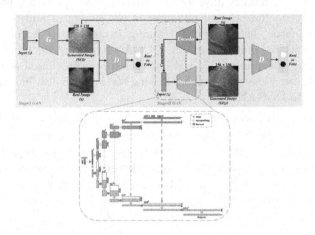

Fig. 1. Overview of the proposed GAN-based network architecture. Stage-I GAN produces mid-resolution images and Stage-II GAN outputs high-resolution images with realistic tissue details. Dashed vertical lines represent skip-connection. Bottom: encoder-decoder architecture to integrate features.

2.1   Stage-I GAN

GAN [9], as an unsupervised generative model to learn the data distribution, is made of two distinct models: a generator G to generate samples as realistic as possible and a discriminator D to discriminate the belonging of the given sample. G aims to transform a latent space vector $z \sim p(z)$ sampled from a prior distribution into a real-like image, while D learns to distinguish between the real image x and the fake image $G(z)$. GAN is trained by minimizing the following adversarial loss in an alternating manner [28], which falls in a state of the confrontational game:

$$L_D = -\mathbb{E}_{x \sim p_{data}(x)}[logD(x)] - \mathbb{E}_{z \sim p_z(z)}[log(1 - D(G(Z)))],$$
$$L_G = -\mathbb{E}_{z \sim p_z(z)}[logD(G(Z))]$$

$$(1)$$

In this work, for Stage-I GAN we adopt several popular GAN-based architectures to synthesize NAFLD US images: DCGAN [20], DCGAN with spectral normalization (SN) [28], and DCGAN with SN [18] and self-attention module (SA) [28].

DCGAN: DCGAN [20] introduces CNN into the generative model to acquire the powerful feature extraction capability. Compared to the traditional GAN design, both the discriminator and the generator in DCGAN discard pooling layers and choose to use convolutional and convolutional-transpose layers respectively. The LeakyReLU activation is utilized in all layers of the discriminator to prevent gradient sparseness and the output of the last convolutional layer is processed by the Sigmoid function neither fully connected layer to give a discriminative result. In the generator, the output layer uses the activation function of $Tanh$ and the remaining layers use ReLU activations. Batch normalization in networks (not include the output layer of the generator and the input layer of the discriminator) helps prevent training issues caused by poor initialization. For uncoditional image synthesis, DCGAN has been widely adopted in the medical imaging community [27] and has therefore been chosen as our baseline GAN architecture. However, in the following sections we also introduce further improvements to DCGAN.

Spectral Normalization (SN) Module: The main idea of SN is to restrict the output of each layer through the Lipschitz constant without complicated parameter adjustments [18]. Doing so can constrain the update of the discriminator triggered by generator update to a lesser extent. The normalization is applied in the generator and discriminator simultaneously. Most recently SN was incorporated into GAN for improving low dose chest X-ray image resolution [26] and multi-modal neuroimage synthesis [13].

Self-Attention (SA) Module: SA module calculates the attention value between local pixel regions and helps to model global correlation in a wider range. The generator with the SA module learns specific structure and geometric features [28]. In addition, the discriminator can now perform complex geometric constraints more accurately on the global structure. This module was recently incorporated into a GAN architecture for synthesizing bone US data [3].

2.2 Stage-II GAN

Directly generating high-resolution images usually meets problems of detail and poor diversity. Instead, we turn to generate a mid-resolution image with high quality in Stage-I and obtain its feature maps in different depth, which are fused into corresponding layers of the generator for information supplement. The generated mid-resolution images have relatively diverse feature distributions but lack vivid tissue representations. The latter generator is supported by rich structure information from synthetic images with the size of 256 × 256 and fills feature details at the same time.

The generated image in Stage-I GAN is fed into the Stage-II GAN generator. The Stage-II GAN generator is constructed using an encoder and decoder network architecture (Fig. 1). The encoder receives Stage-I images and outputs feature maps in various sizes. The basic block in the encoder is comprised of a convolutional layer and a maxpooling layer. The downsampling enlarges the receptive field area and concentrates on feature extraction. Captured features are integrated into the generator by skip-connection. We also concatenate the random noise vector z in the encoder output and input this combined feature vector to the decoder. Conditioned on the low-resolution result, obtained Stage-I, and the noise vector the discriminator and generator of Stage-II GAN are trained by maximizing L_D and minimizing L_G showed in Eq. 1. The proposed method avoids the prior knowledge from real data as input and guarantees the diversity of generated images. Our discriminator, denoted as D, during this stage uses the same architecture of discriminator in DCGAN to perform differentiation of real or synthetic. Using information from generated mid-resolution images rather than real images prevents the model from memorizing patterns from real images. Furthermore, Stage-II GAN corrects imaging artifacts in the low-resolution image, obtained in Stage-I, synthesizing high-resolution realistic liver US data.

3   Experiments and Results

3.1   Dataset

Experiments are performed on the dataset provided by [6]. The NAFLD dataset includes 550 B-mode US scans and biopsy results from 55 subjects. 10 US images were collected for each subject. Using biopsy, 38 subjects were diagnosed as NAFLD patients and the rest 17 were viewed as normal/healthy individuals. The data was collected using the GE Vivid E9 Ultrasound System (GE Healthcare INC, Horten, Norway) equipped with a sector US transducer operating at 2.5 MHz [6]. All images were cropped to remove irrelevant regions (mostly related to text involving image acquisition settings) and then resized to a size of 256×256. During the cropping, the original image resolution of 0.373 mm was kept constant.

Training and Test Data: To validate the performance of our proposed method, we randomly split 70 normal and 70 diseased images as a testing set, remaining images (100 healthy and 310 diseased) are grouped as a training set to train GAN-based networks. The classification network was trained using real and synthesized data totaling to 1000 healthy and diseased US images. The split obeys the rule that the same patient scans are not used for both training and testing. The random split operation was repeated five time and average results are reported.

We conduct experiments using the PyTorch framework with an Intel Core CPU at 3.70 GHz and an Nvidia GeForce GTX 1080Ti GPU. GAN-based networks are trained using the cross-entropy loss and ADAM optimization method with batch size of 16 and a learning rate of 0.0002. Exponential decay rate for the first and second moment estimates are set to 0.5 and 0.999. The performance of our proposed generative method is compared with those popular GAN-based architectures mentioned in Sect. 2.1 to generate liver US images directly. Examples for each class are generated individually, not incorporating class conditions.

3.2   GAN-Based Network Evaluation

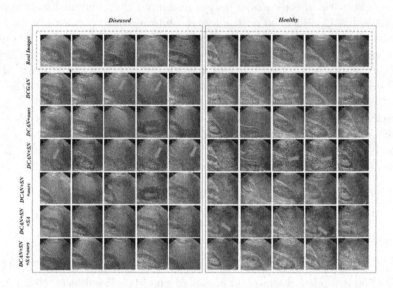

Fig. 2. Qualitative results of the images generated by DCGAN in combination with the different modules, along with the proposed module. SA - Self-attention module and SN - Spectral normalization module. Blue arrows point to qualitative improvements achieved over prior state of the art. In all the presented results *ours* denotes the integration of Stage-II GAN module.

Qualitative Results: Qualitative results for the investigated models are given in Fig. 2. The first row in the figure demonstrates examples of real images followed by the synthesized images generated by different methods, for both classes diseased and healthy. The different state of the art methods that are used to synthesize the images and compare them are, DCGAN [20], DCGAN combined with the proposed (Stage-II GAN) module, DCGAN [20] with SN [18] module, DCGAN [20] with SN [18] module combined with the proposed (Stage-II GAN) module, DCGAN [20] with SN [18] and SA [28] modules, and a combination of DCGAN [20] with all modules (SN [18], SA [28] and proposed Stage-II GAN)). Investigating the results in Fig. 2 it can be seen that when the proposed module (Stage-II GAN) is used in combination with state of art DCGAN modules, qualitative improvements can be obtained. Blue arrows in Fig. 2 point to anatomy missed using the prior GANs investigated. However, we can see that by incorporating our proposed module liver tissue characterization in the synthesized images improves.

Quantitative Results: The *Inception Score* (IS) and *Frechet Inception Distance* (FID) score are used to quantitatively evaluate the generated image quality and diversity. The *Inception Score* (IS) helps to estimate the quality of the generated images based on the classification performance of Inception V3 classifier on the synthesized images [10]. Higher IS value means the synthesized images are diverse and similar to the real data [11, 27]. Although IS is a very good metric to assess the quality of the synthesized images, it does not compare the synthetic images with the original images. *Frechet Inception Distance* (FID) is based on the statistics of the generated images compared with that of the original images [10]. Similar to IS, FID is also calculated using the Inception V3 model, the activations of the last pooling layer are summarized as a multi-variate Gaussian, the distance between the two Gaussians are calculated as FID [10]. A low FID shows that the images synthesized by this GAN architecture have high diversity in them and are at par with the real images [11, 27].

The IS and FID metrics are calculated on 400 synthesized images for each category. From Table 1, it is noted that our Stage-II GAN module significantly improved the IS and FID results for all the investigated prior GAN modules (paired t-test $p < 0.05$). The highest IS score is obtained when DCGAN is combined with the SN and our proposed Stage-II GAN module. The lowest FID score is achieved when DCGAN is combined with the proposed model, that is without the SA and SN modules.

Table 1. IS and FID of the proposed DCGAN, DCGAN+SN, DCGAN+SN+SA to synthesize liver US images directly and incorporating our Stage-II GAN. Bold text shows the best results obtained. SA- Self attention module, SN- Spectral Normalization module. In all the presented results *ours* denotes the integration of Stage-II GAN module.

	IS↑ abnormal/normal	FID ↓ abnormal/normal
DCGAN [20]	$1.32 \pm 0.02/1.28 \pm 0.01$	113.87/161.76
DCGAN [20] + ours	$1.55 \pm 0.08/1.48 \pm 0.05$	**100.05/99.53**
DCGAN [20] + SN [18]	$1.09 \pm 0.01/1.34 \pm 0.06$	170.68/247.76
DCGAN [20] + SN [18] + ours	$\mathbf{1.67 \pm 0.08}/1.50 \pm 0.05$	156.55/110.17
DCGAN [20] + SN [18] + SA [28]	$1.42 \pm 0.03/1.38 \pm 0.03$	160.19/259.19
DCGAN [20] + SN [18] + SA [28] + ours	$1.51 \pm 0.07/\mathbf{1.51 \pm 0.06}$	108.39/103.07

3.3 Classification Evaluation

To evaluate the quality of the synthesized images, EfficientNet [23] is employed to perform binary classification on the original dataset and expanded class-balanced dataset. The training dataset is expanded from 410 images to 1000 images using 590 generated images (Diseased class: 310 real + 190 synthetic; Healthy class: 100 real + 400 synthetic). As explained previously, test data was 140 real US images (70 health 70 diseased) which were not part of the image synthesis process. The classification performance is measured by *accuracy, precision, recall* and $F1_{score}$. The quantitative results are shown in Table 2. From the table, it can be noted that the classification algorithm obtains the best accuracy when the synthesized images are obtained using DCGAN in combination with the proposed Stage-II GAN module (paired t-test $p < 0.05$). Similar to GAN evaluation results, from Table 2 we can observe that our proposed Stage-II GAN module significantly improves classification performance metrics for all the investigated prior GAN modules (paired t-test $p < 0.05$).

Table 2. Quantitative classification results for all the investigated methods. Bold text shows the best results obtained. In all the presented results *ours* denotes the integration of Stage-II GAN module.

	Accuracy	Precision	Recall	$F1_{score}$
The original dataset	82.14%	82.47%	82.14%	82.10%
DCGAN [20]	84.29%	84.31%	84.29%	84.29%
DCGAN [20] + ours	**85.71%**	**87.68%**	**85.71%**	**85.53%**
DCGAN [20] + SN [18]	74.29%	76.00%	74.29%	73.85%
DCGAN [20] + SN [18] + ours	78.57%	79.17%	78.57%	78.46%
DCGAN [20] + SN [18] + SA [18]	80.71%	80.77%	80.71%	80.71%
DCGAN [20] + SN [18] + SA [18] + ours	82.86%	82.88%	82.86%	82.85%

4 Conclusion

In this work, a novel GAN architecture for realistic B-mode liver US image generation was proposed. Qualitative and quantitative results show significant improvements in image synthesis can be achieved using the proposed two-stage architecture. We also show that the classification performance of well-known CNN architectures can be significantly improved using the synthesized images. Our study is the first attempt to synthesize diseased and healthy liver US images based on a novel GAN module that can be incorporated into popular GAN-based models for improving their performance. One major drawback of our work is the limited dataset size. We only had access to 550 B-mode US data. Increasing the dataset size could also result in the performance improvements of the classification method investigated in this work. Furthermore, we have only evaluated the performance of DCGAN as Stage-I GAN architecture. Investigation of various other GAN architectures, used for medical image synthesis [11,27], should also be performed to understand the full potential of our Stage-II GAN model. Finally, a comparison study against traditional augmentation methods should also be performed. Future work will include the collection of large-scale liver US data and improvements of the shortcomings of our work.

References

1. Acharya, U.R., et al.: Automated characterization of fatty liver disease and cirrhosis using curvelet transform and entropy features extracted from ultrasound images. Comput. Biol. Med. **79**, 250–258 (2016)
2. Ali, I.S., Mohamed, M.F., Mahdy, Y.B.: Data augmentation for skin lesion using self-attention based progressive generative adversarial network. Exp. Syst. Appl. **165**, 113922 (2019)
3. Alsinan, A.Z., Rule, C., Vives, M., Patel, V.M., Hacihaliloglu, I.: GAN-based realistic bone ultrasound image and label synthesis for improved segmentation. In: Martel, A.L., et al. (eds.) MICCAI 2020. LNCS, vol. 12266, pp. 795–804. Springer, Cham (2020). https://doi.org/10.1007/978-3-030-59725-2_77
4. Amarapurkar, D., et al.: Prevalence of non-alcoholic fatty liver disease: population based study. Ann. Hepatol. **6**(3), 161–163 (2007)
5. Biswas, M., et al.: Symtosis: a liver ultrasound tissue characterization and risk stratification in optimized deep learning paradigm. Comput. Methods Prog. Biomed. **155**, 165–177 (2018)
6. Byra, M., et al.: Transfer learning with deep convolutional neural network for liver steatosis assessment in ultrasound images. Int. J. Comput. Assist. Radiol. Surg. **13**(12), 1895–1903 (2018)
7. Che, H., Brown, L.G., Foran, D.J., Nosher, J.L., Hacihaliloglu, I.: Liver disease classification from ultrasound using multi-scale CNN. Int. J. Comput. Assist. Radiol. Surg. **16**, 1537-1548 (2021)
8. Gaidos, J.K., Hillner, B.E., Sanyal, A.J.: A decision analysis study of the value of a liver biopsy in nonalcoholic steatohepatitis. Liver Int. **28**(5), 650–658 (2008)
9. Goodfellow, I., et al.: Generative adversarial networks. Commun. ACM **63**(11), 139–144 (2020)
10. Heusel, M., Ramsauer, H., Unterthiner, T., Nessler, B., Hochreiter, S.: Gans trained by a two time-scale update rule converge to a local nash equilibrium. arXiv preprint arXiv:1706.08500 (2017)
11. Kazeminia, S., et al..: Gans for medical image analysis. Artif. Intell. Med. **109**, 101938 (2020)
12. Khov, N., Sharma, A., Riley, T.R.: Bedside ultrasound in the diagnosis of nonalcoholic fatty liver disease. World J. Stroenterol. WJG **20**(22), 6821 (2014)
13. Lan, H., Toga, A.W., Sepehrband, F., Initiative, A.D.N., et al.: SC-GAN: 3D self-attention conditional GAN with spectral normalization for multi-modal neuroimaging synthesis. bioRxiv (2020)
14. Li, Q., Dhyani, M., Grajo, J.R., Sirlin, C., Samir, A.E.: Current status of imaging in nonalcoholic fatty liver disease. World J. Hepatol. **10**(8), 530 (2018)
15. Liu, X., Song, J.L., Wang, S.H., Zhao, J.W., Chen, Y.Q.: Learning to diagnose cirrhosis with liver capsule guided ultrasound image classification. Sensors (Basel, Switzerland) **17**(1), 149 (2017)
16. Meng, D., Zhang, L., Cao, G., Cao, W., Zhang, G., Hu, B.: Liver fibrosis classification based on transfer learning and fcnet for ultrasound images. IEEE Access **5**, 5804–5810 (2017)
17. Milletari, F., Navab, N., Ahmadi, S.A.: V-net: Fully convolutional neural networks for volumetric medical image segmentation (2016)
18. Miyato, T., Kataoka, T., Koyama, M., Yoshida, Y.: Spectral normalization for generative adversarial networks. arXiv preprint arXiv:1802.05957 (2018)

19. Nasr, P., Ignatova, S., Kechagias, S., Ekstedt, M.: Natural history of nonalcoholic fatty liver disease: a prospective follow-up study with serial biopsies. Hepatol. Commun. **2**(2), 199–210 (2018)
20. Radford, A., Metz, L., Chintala, S.: Unsupervised representation learning with deep convolutional generative adversarial networks. arXiv preprint arXiv:1511.06434 (2015)
21. Reddy, D.S., Bharath, R., Rajalakshmi, P.: Classification of nonalcoholic fatty liver texture using convolution neural networks. In: 2018 IEEE 20th International Conference on e-Health Networking, Applications and Services (Healthcom), pp. 1–5 (2018)
22. Strauss, S., Gavish, E., Gottlieb, P., Katsnelson, L.: Interobserver and intraobserver variability in the sonographic assessment of fatty liver. Am. J. Roentgenol. **189**(6), W320–W323 (2007)
23. Tan, M., Le, Q.: Efficientnet: rethinking model scaling for convolutional neural networks. In: International Conference on Machine Learning, pp. 6105–6114. PMLR (2019)
24. Tapper, E.B., Lok, A.S.F.: Use of liver imaging and biopsy in clinical practice. New Engl. J. Med. **377**(8), 756–768 (2017)
25. Targher, G., Day, C.P., Bonora, E.: Risk of cardiovascular disease in patients with nonalcoholic fatty liver disease. New Engl. J. Med. **363**(14), 1341–1350 (2010)
26. Xu, L., Zeng, X., Huang, Z., Li, W., Zhang, H.: Low-dose chest x-ray image super-resolution using generative adversarial nets with spectral normalization. Biomed. Sig. Process. Control **55**, 101600 (2020)
27. Yi, X., Walia, E., Babyn, P.: Generative adversarial network in medical imaging: a review. Med. Image Anal. **58**, 101552 (2019)
28. Zhang, H., Goodfellow, I., Metaxas, D., Odena, A.: Self-attention generative adversarial networks. In: International Conference on Machine Learning, pp. 7354–7363. PMLR (2019)

Quality Assessment and Quantitative Imaging

Adaptable Image Quality Assessment Using Meta-Reinforcement Learning of Task Amenability

Shaheer U. Saeed[1(✉)], Yunguan Fu[1,2], Vasilis Stavrinides[3,4],
Zachary M. C. Baum[1], Qianye Yang[1], Mirabela Rusu[5], Richard E. Fan[6],
Geoffrey A. Sonn[5,6], J. Alison Noble[7], Dean C. Barratt[1], and Yipeng Hu[1,7]

[1] Centre for Medical Image Computing, Wellcome/EPSRC Centre for Interventional
and Surgical Sciences, and Department of Medical Physics and Biomedical
Engineering, University College London, London, UK
shaheer.saeed.17@ucl.ac.uk
[2] InstaDeep, London, UK
[3] Division of Surgery and Interventional Science, University College London,
London, UK
[4] Department of Urology, University College Hospital NHS Foundation Trust,
London, UK
[5] Department of Radiology, Stanford School of Medicine, Stanford, CA, USA
[6] Department of Urology, Stanford School of Medicine, Stanford, CA, USA
[7] Department of Engineering Science, University of Oxford, Oxford, UK

Abstract. The performance of many medical image analysis tasks are
strongly associated with image data quality. When developing modern
deep learning algorithms, rather than relying on subjective (human-
based) image quality assessment (IQA), task amenability potentially provides an objective measure of task-specific image quality. To predict task
amenability, an *IQA agent* is trained using reinforcement learning (RL)
with a simultaneously optimised *task predictor*, such as a classification or
segmentation neural network. In this work, we develop transfer learning
or adaptation strategies to increase the adaptability of both the IQA
agent and the task predictor so that they are less dependent on high-
quality, expert-labelled training data. The proposed transfer learning
strategy re-formulates the original RL problem for task amenability in a
meta-reinforcement learning (meta-RL) framework. The resulting algorithm facilitates efficient adaptation of the agent to different definitions
of image quality, each with its own Markov decision process environment including different images, labels and an adaptable task predictor.
Our work demonstrates that the IQA agents pre-trained on non-expert
task labels can be adapted to predict task amenability as defined by
expert task labels, using only a small set of expert labels. Using 6644
clinical ultrasound images from 249 prostate cancer patients, our results
for image classification and segmentation tasks show that the proposed
IQA method can be adapted using data with as few as respective 19.7%
and 29.6% expert-reviewed consensus labels and still achieve comparable IQA and task performance, which would otherwise require a training
dataset with 100% expert labels.

© Springer Nature Switzerland AG 2021
J. A. Noble et al. (Eds.): ASMUS 2021, LNCS 12967, pp. 191–201, 2021.
https://doi.org/10.1007/978-3-030-87583-1_19

Keywords: Image quality assessment · Meta-reinforcement learning ·
Task amenability · Ultrasound · Prostate

1 Introduction

Medical image quality can influence the downstream clinical tasks intended for
medical images [1]. Automated algorithms have been proposed for image quality
assessment (IQA), based on human scoring of image quality [2–6], prior clinical
knowledge [7,8] or a set of hand-engineered criteria [9–11]. Task-specific image
quality, which measures how well a clinical task can be completed using the
image being assessed, may be preferred, but previous methods still rely on human
interpretation [2,3]. When the downstream clinical tasks are completed by automated machine learning algorithms, task-specific IQA may become more relevant, however, human perceived task-specific IQA may not accurately reflect the
performance of the machine optimised *task predictors*. Recent works introduce
task amenability; defined as the task-specific image quality to directly measure
target task performance [12,13], which also takes into account the dependency
between training an automated IQA and the training of a task predictor.

For predicting task amenability for IQA, Saeed *et al.* [12] proposed to train a
controller; here, a reinforcement learning (RL) agent, together with the task predictor. Classification and segmentation neural networks were tested as the task
predictors. The trained controller predicts significantly different task amenability scores to those determined by humans, with or without requiring human
labels of task amenability during training.

By definition, this IQA approach is inevitably dependent on the task predictor and the labelled data used to train such a task predictor, in the case of
supervised learning. In clinical practice, the feasibility and cost associated with
obtaining quality labelled data sets for various target tasks can not be overlooked.
Therefore, we propose a transfer learning strategy to train the IQA agent based
on meta-reinforcement learning (meta-RL) across multiple environments. These
RL environments can then be designed to reflect different Markov decision processes (MDPs) with differently labelled data. At the same time, a shared task
predictor[1] is trained between these MDPs, such that it may be adapted together
with the meta-trained controller. Equipping adaptation ability to both the controller and the task predictor has several potential applications for the efficient
use of labelled data. In this work, we demonstrate the resulting adaptation ability from relatively low-quality *non-expert* task labels annotated by individual
observers to high-quality *expert* labels carefully curated by reviewed consensus.

The contributions of the work are summarised as follows: 1) we propose
a transfer learning or adaptation strategy to train an adaptable IQA system;
2) we design a meta-RL algorithm for training the task-amenability-predicting
controller together with a target task predictor, which is shared amongst multiple environments, such that training to convergence is not required on every

---

[1] *Tasks* refer to the target classification or segmentation tasks, while MDPs or environments are preferred over *meta-tasks* found in meta-learning literature for clarity.

time-step and where adaptability is equipped to both the inner and outer loops simultaneously; 3) we demonstrate the efficacy of the proposed transfer learning strategy with experiments using a large set of clinical ultrasound images from prostate cancer patients, labelled by four different observers with varying experience and expertise; 4) the experiments show that using 20–30% of the expert labels is sufficient to fine-tune both the RL controller and the task predictor to achieve comparable performances to when they are trained using the full set of expert labels.

2 Methods

2.1 Image Quality Assessment by Task Amenable Data Selection

In this work, we follow the IQA formulation proposed by Saeed *et al.* [12]. There are two parametric functions, a task predictor $f(\cdot; w) : \mathcal{X} \to \mathcal{Y}$ and a controller $h(\cdot; \theta) : \mathcal{X} \to [0, 1]$, with parameters w and θ, respectively. $\mathcal{X}$ and $\mathcal{Y}$ are the respective image and label domains with $\mathcal{P}_{XY}$ being the joint image-label distribution, with a density function $p(x, y)$.

The task predictor f is optimised to predict labels, by minimising the loss function $L_f : \mathcal{Y} \times \mathcal{Y} \to \mathbb{R}_{\geq 0}$ using sampled data:

$$\min_{w} \mathbb{E}_{(x,y) \sim \mathcal{P}_{XY}^h} [L_f(f(x; w), y)], \tag{1}$$

where $\mathcal{P}_{XY}^h$ is the controller-selected joint image-label distribution, with density function $p^h(x, y) \propto p(x, y) h(x; \theta)$.

The controller h is optimised to measure image quality (task amenability), by minimising the metric function $L_h : \mathcal{Y} \times \mathcal{Y} \to \mathbb{R}_{\geq 0}$:

$$\min_{\theta} \mathbb{E}_{(x,y) \sim \mathcal{P}_{XY}^h} [L_h(f(x; w), y)] \tag{2}$$

where L_h is in general a non-differentiable metric computed on the validation set, and different to L_f.

The optimisation is performed using reinforcement learning, where the environment consists of the training set from $\mathcal{P}_{XY}$ and the task predictor $f(\cdot; w)$; the agent is the controller $h(\cdot; \theta)$ whose action is sample selection $a_t = \{a_{i,t}\}_{i=1}^{B} \in \{0, 1\}^B$, based on the predicted quality scores $\{h(x_i; \theta)\}_{i=1}^{B}$, from a mini-batch of training samples $\mathcal{B}_t = \{(x_i, y_i)\}_{i=1}^{B}$; and the reward is the task predictor performance on a validation set from the same distribution $\mathcal{P}_{XY}$, which is computed after training, for a fixed number of steps, using the selected samples. In this work we use the reward formulation, from [12], which does not require human task amenability labels, and weights the validation set using controller predictions. R_t is thus the reward which is a weighted sum of validation set performance.

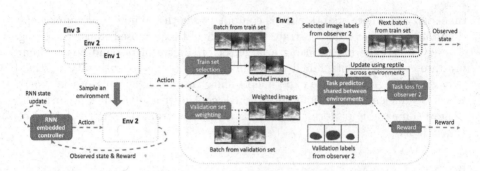

Fig. 1. An overview of the proposed meta-RL framework for training the task predictor and the RNN-embedded controller (the IQA agent).

2.2   Meta-Reinforcement Learning with Different Labels

In this section, we consider multiple label distributions $\{\mathcal{P}^k_{Y|X}\}^K_{k=1}$, such that each sample x has multiple labels $\{y^k\}^K_{k=1}$. The joint distributions are thereby $\mathcal{P}^k_{XY} = \mathcal{P}_X \mathcal{P}^k_{Y|X}$ for $k = 1, \ldots, K$. Each distribution $\mathcal{P}^k_{XY}$ forms an RL environment with an MDP M_k. These MDPs are assumed to be sampled from the same MDP distribution $\mathcal{P}_M$, i.e. $M_k \sim \mathcal{P}_M$. The task predictors $f(\cdot; w)$ and controller $h(\cdot; \theta)$ are both shared across different environments.

We adopt the meta-RL formulation [14,15] for reinforcement learning across multiple environments. Given a set of MDPs $\{M_k\}^K_{k=1}$, a *trial* is defined as multiple episodes with a sampled MDP M_k. The meta-RL agent learns across multiple trials by sampling $M_k \sim \mathcal{P}_M$. Different from the RL with one single fixed environment, at time $t + 1$, the meta-RL agent h takes the action a_t, raw reward r_t, and termination flag d_t at the previous time step in addition to the observed current state s_{t+1}. Note that for per-sample operation $r_t = R_t$ at the episode end, and zero otherwise, similar to sparse reward formulations in [14,15]. Denote the input tuple as $\tau_{t+1} = (s_{t+1}, a_t, r_t, d_t)$, thereby $h(\cdot; \theta)$ is now defined with a space of $\mathcal{X} \times [0, 1] \times \mathbb{R} \times \{0, 1\}$.

In this work, the meta-RL agent adopts a recurrent neural network (RNN) with internal memory shared across episodes in the same trial. Importantly, the internal memory is reset when a trial finishes, i.e. before another environment is sampled. This mechanism allows test-time adaptability, even with fixed weights [16–20], and thereby transfers knowledge between environments [14,15,21]. This is due to the RNN making the controller a function of the history leading up to a sample such that changing history can influence the action for that sample. The full algorithm is described in Algorithm 1, with details for configuring episodic mini-batches and meta-loop trials. An overview is also presented in Fig. 1. In our implementation, proximal policy optimisation (PPO) [22] was used to train the controller. The task predictor employs the Reptile scheme [23] to allow potential data efficiency benefit for adapting to different observer labels. The predictor is updated in two steps: 1) update starting weights w_{t+1} of predictor $f(\cdot; w_{t+1})$

to $w_{t+1,\text{new}}$, using gradient descent based on $\mathcal{B}_{t,\text{selected}}$; 2) update weights using $w_{t+1} \leftarrow w_{t+1} + \epsilon(w_{t+1,\text{new}} - w_{t+1})$ where ϵ is 1.0 initially and is linearly annealed to 0.0 as trial iterates. It is worth noting that the IQA algorithm from [12] can be considered a special case of our proposed method with only one environment.

After training using the scheme described in Algorithm 1, the adaptation stage, for both the controller and task-predictor, can be performed on a single MDP of interest $M_a \sim \mathcal{P}_M$, where M_a is the environment which we would like to adapt to. If multiple iterations of the outer loop are required, the internal state of the controller is only reset on the first iteration. The controller weights remain fixed; adaptability is a result of updating internal state.

Algorithm 1: Adaptable image quality assessment by task amenability

---

Data: Multiple MDPs $M_k \sim \mathcal{P}_M$.
Result: Task predictor $f(\cdot; w)$ and controller $h(\cdot; \theta)$.

while *not converged* **do**
  Sample an MDP $M_k \sim \mathcal{P}_M$;
  Reset the internal state of controller h;
  for *Each episode in all episodes* **do**
    for $t \leftarrow 1$ **to** T **do**
      Sample a training mini-batch $\mathcal{B}_t = \{(x_{i,t}, y_{i,t})\}_{i=1}^{B}$;
      Compute selection probabilities $\{h_{i,t}\}_{i=1}^{B} = \{h(\tau_{i,t}; \theta_t)\}_{i=1}^{B}$;
      Sample actions $a_t = \{a_{i,t}\}_{i=1}^{B}$ w.r.t. $a_{i,t} \sim \text{Bernoulli}(h_{i,t})$;
      Select samples $\mathcal{B}_{t,\text{selected}}$ from $\mathcal{B}_t$;
      Update predictor $f(\cdot; w_t)$ with $\mathcal{B}_{t,\text{selected}}$ using Reptile;
      Compute reward R_t;
    end
    Collect one episode $\{\mathcal{B}_t, a_t, R_t\}_{t=1}^{T}$;
    Update controller $h(\cdot; \theta)$ using the RL algorithm PPO;
  end
end

---

3   Experiments

In this work we use 6644 2D ultrasound images from 249 prostate cancer patients. During the early stages of ultrasound-guided biopsy procedures, images were acquired using a transperineal ultrasound probe (C41L47RP, HI-VISION Preirus, Hitachi Medical Systems Europe) as part of SmartTarget: THER-APY and SmartTarget: BIOPSY clinical trials (clinicaltrials.gov identifiers NCT02290561 and NCT02341677 respectively). Images from each subject initially consisted of 50–120 frames. For feasibility of manual labelling, frames were sampled at four-degree intervals where relative rotation angles were tracked using a digital transperineal stepper (D&K Technologies GmbH, Barum, Germany). The resulting 6644 2D ultrasound images were randomly split, at the patient level, into training, validation and holdout sets, with 4429, 1092 and 1123 images from 174, 37 and 38 subjects, respectively.

Three sets of task label $\{L_i\}_{i=1}^3$ were collected from three trained biomedical engineering researchers. These individually-labelled are referred to as "non-expert" label sets for brevity. In addition, the fourth set of "expert" labels L_* was curated by a urologist, first carefully reviewing a reference set of consensus labels and then editing them as deemed necessary. For all label sets, each image has both a binary label indicating prostate presence for classification and a binary mask of the prostate gland for segmentation.

The task predictor algorithms used for the two tested applications are the same as [12]. For classification, AlexNet [24] was used with a cross-entropy loss and a reward based on classification accuracy. For segmentation, U-Net [25] was used with a pixel-wise cross-entropy loss and a reward based on mean binary Dice score. The controller had a three-layer convolutional encoder, before feeding the encoded features to an RNN with a stacked-LSTM architecture, as described in [15]. Experimental results are reported for empirically configured networks and default hyper-parameter values remain unchanged unless specified.

The following three different IQA models were trained and compared.

- *Baseline*: Trained with all training and validation data using only the high-quality expert labels L_*. That is, only one "expert-labelled" environment in training, establishing a reference for achievable IQA system performance.
- *Meta-RL*: The proposed model that was first trained with training and validation data using the non-expert labels $\{L_i\}_{i=1}^3$ as three different environments. Both the task predictor and the controller were subsequently adapted with $k \times 100\%$ training and validation data using the expert labels L_*.
- *Meta-RL Variant*: For comparison, a basic implementation of transfer learning. The model was first trained with all training and validation data using the shuffled non-expert labels $\{L_i\}_{i=1}^3$ as one single environment, i.e. without considering different environment-specific trials, and the Reptile update for optimising the task predictor reduced to standard gradient descent. Adaptation was done with $k \times 100\%$ training and validation data using the expert labels L_*. The internal state of RNN was not reset before fine-tuning.

We evaluate the IQA models jointly with the task predictors using task performance, which serves as both a direct evaluation of the task-predictor and an indirect evaluation of the IQA agent by its task amenability definition. We report mean accuracy (Acc.) and mean binary Dice score (Dice) on the holdout set using expert labels for classification and segmentation, respectively. These measures are averaged over all 2D slices in the holdout set. Where controller selection is used, the metric is computed over the selected samples only. Samples are selected by rejecting the subset with the lowest controller predicted values, with the specified rejection ratios. Standard deviation (St.D) is reported to measure inter-patient variance, with which, paired t-test results with a significance level of 5% are reported when any comparison is made. We evaluate the models for varying k-values, where k is the ratio of expert-labelled samples used for adaptation ($k \times 100\%$ samples used).

4 Results

Table 1. Comparison of holdout set results with a rejection ratio set to 5%

Tasks		Prostate classification (Acc.)	Prostate segmentation (Dice)
IQA Methods	k	Mean ± St.D.	Mean ± St.D.
Baseline	N/A	0.932 ± 0.011	0.894 ± 0.016
Meta-RL	0.5	0.936 ± 0.012	0.892 ± 0.018
	0.4	0.929 ± 0.016	0.886 ± 0.014
	0.3	0.926 ± 0.010	0.888 ± 0.020
	0.2	0.925 ± 0.017	0.873 ± 0.017
	0.1	0.911 ± 0.012	0.863 ± 0.020
	0.0	0.908 ± 0.010	0.857 ± 0.018
Meta-RL Variant	0.5	0.931 ± 0.015	0.884 ± 0.016
	0.4	0.920 ± 0.010	0.882 ± 0.021
	0.3	0.919 ± 0.013	0.882 ± 0.015
	0.2	0.916 ± 0.014	0.860 ± 0.014
	0.1	0.905 ± 0.014	0.858 ± 0.021
	0.0	0.896 ± 0.016	0.849 ± 0.017

The proposed meta-training took, on average, approximately 48 h and the meta-testing (model fine-tuning) took 1–2 h on a single Nvidia Quadro P5000 GPU. This result reflects the design of the proposed adaptation strategy for data efficiency and, arguably, also for computational efficiency.

Performance of the IQA models, in terms of Acc. and Dice, are summarised in Table 1 and plotted in Fig. 2 against varying k values. In the prostate presence classification task, no statistical significance was found between the baseline and meta-RL for k values from 0.5 to 0.2 (p-values ranged from 0.10 to 0.23). However, meta-RL performance for low k values, $k = 0.1$ or 0.0, was significantly lower than that of the baseline ($p < 0.01$ for both). In the prostate segmentation task, no statistical significance was found between the two, for k-values from 0.5 to 0.3 (p-values ranged from 0.07 to 0.17), but a significantly lower performance was found for meta-RL for low k values from 0.2 to 0.0 ($p < 0.01$ for all).

For the ablation study comparing meta-RL to the meta-RL variant, the proposed meta-RL framework generally outperformed the meta-RL variant for the same k values, for both tested target tasks, as detailed in Table 1. For classification, we report a statistically significant difference between the two, for the same k values from 0.0 to 0.4 ($p < 0.01$ for all), while no significance was found when the k increased to 0.5 ($p = 0.06$). For segmentation, superior performance from the proposed meta-RL was statistically significant for all k values ($p < 0.03$ for all). From an ablation study, with and without the Reptile scheme for updating task predictors, the Reptile-omitted meta-RL classification and segmentation tasks achieved Acc. = 0.901 ± 0.013 and Dice = 0.851 ± 0.013, respectively, when

$k = 0$. The improvement, when using the Reptile scheme, was statistically significant with $p < 0.01$ for both, but no significant difference was found for other k values.

Figure 3 provides visual examples of selection decisions by the adapted IQA agent. With 5% rejection ratio, all these rejected examples seem visually challenging for respective classification and segmentation tasks, and rejecting these examples improved performances of the simultaneously learned task predictors.

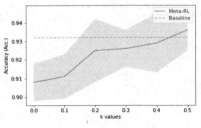

(a) Prostate presence classification task

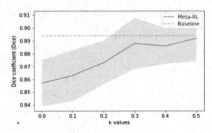

(b) Prostate segmentation task

Fig. 2. Task performance (in respective Acc. and Dice metrics) against the k values with rejection ratio set to 5%.

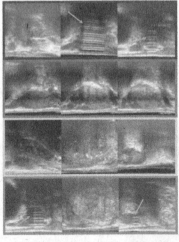

(a) Prostate classification task

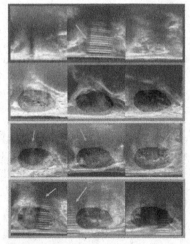

(b) Prostate segmentation task

Fig. 3. Examples of controller selected and rejected images (rejection ratio = 5%) for both tasks. **Blue:** rejected samples; **Red:** selected samples; **Yellow:** rejected samples despite no apparent artefacts or severe noise; **Green:** selected samples despite present artefacts or low contrast. **Orange arrows:** visible artefacts; **Cyan arrows:** regions where gland boundary delineation may be challenging. (Color figure online)

5   Discussion and Conclusion

Based on results reported in Sect. 4, for the tested ultrasound guidance application, the proposed adaptation strategy allows for the IQA agent and task predictor to be adapted using as few as 1087 and 1634 expert-labelled images from 42 and 63 subjects (training and validation sets), for classification and segmentation, respectively. Compared with a total of 5521 expert-labelled images from 211 subjects that were required to train the baseline, this is a substantial reduction, to 20–30%, in the required quantity of high-quality and often expensive expert-labelled data. The proposed model also used non-expert labels for training but these may be used for different IQA definitions, further economic analysis is beyond the scope of this work. An adaptable IQA algorithm has been presented, which can be efficiently adapted with new labelled data. The proposed algorithm may have general applicability to alleviate demand for large quantities of training data, for example, for other imaging protocols or target tasks.

Acknowledgements. This work is supported by the Wellcome/EPSRC Centre for Interventional and Surgical Sciences [203145Z/16/Z], the CRUK International Alliance for Cancer Early Detection (ACED) [C28070/A30912; C73666/A31378], EPSRC CDT in i4health [EP/S021930/1], the Departments of Radiology and Urology, Stanford University, an MRC Clinical Research Training Fellowship [MR/S005897/1] (VS), the Natural Sciences and Engineering Research Council of Canada Postgraduate Scholarships-Doctoral Program (ZMCB), the University College London Overseas and Graduate Research Scholarships (ZMCB), GE Blue Sky Award (MR), and the generous philanthropic support of our patients (GAS). Previous support from the European Association of Cancer Research [2018 Travel Fellowship] (VS) and the Alan Turing Institute [EPSRC grant EP/N510129/1] (VS) is also acknowledged.

References

1. Chow, L.S., Paramesran, R.: Review of medical image quality assessment. Biomed. Sig. Process. Control **27**, 145–154 (2016)
2. Esses, S.J., et al.: Automated image quality evaluation of T2-weighted liver MRI utilizing deep learning architecture. J. Magn. Reson. Imag. **47**(3), 723–728 (2018)
3. Zago, G.T., Andreão, R.V., Dorizzi, B., Ottoni, E., Salles, T.: Retinal image quality assessment using deep learning. Comput. Biol. Med. **103**, 64–70 (2018)
4. Baum, Z.M.C., et al.: Image quality assessment for closed-loop computer-assisted lung ultrasound. In: Linte, C.A., Siewerdsen, J.H. (eds.) Medical Imaging 2021: Image-Guided Procedures, Robotic Interventions, and Modeling, vol. 11598, pp. 160–166. International Society for Optics and Photonics. SPIE (2021). https://doi.org/10.1117/12.2581865
5. Abdi, A.H., et al.: Automatic quality assessment of echocardiograms using convolutional neural networks: feasibility on the apical four-chamber view. IEEE Trans. Med. Imag. **36**(6), 1221–1230 (2017). https://doi.org/10.1109/TMI.2017.2690836

6. Liao, Z., et al.: On modelling label uncertainty in deep neural networks: automatic estimation of intra-observer variability in 2D echocardiography quality assessment. IEEE Trans. Med. Imag. **39**(6), 1868–1883 (2019)

7. Wu, L., Cheng, J., Li, S., Lei, B., Wang, T., Ni, D.: FUIQA: fetal ultrasound image quality assessment with deep convolutional networks. IEEE Trans. Cybern. **47**(5), 1336–1349 (2017)

8. Lin, Z., et al.: Multi-task learning for quality assessment of fetal head ultrasound images. Med. Image Anal. 58, 101548 (2019). ISSN: 1361-8415. https://doi.org/10.1016/j.media.2019.101548

9. Davis, H., Russell, S., Barriga, E., Abramoff, M., Soliz, P.: Vision-based, real-time retinal image quality assessment, pp. 1–6 (2009)

10. Köhler, T., Budai, A., Kraus, M.F., Odstrèilik, J., Michelson, G., Hornegger, J.: Automatic no-reference quality assessment for retinal fundus images using vessel segmentation. In: Proceedings of the 26th IEEE International Symposium on Computer-Based Medical Systems, pp. 95–100 (2013)

11. Loizou, C.P., Pattichis, C.S., Pantziaris, M., Tyllis, T., Nicolaides, A.: Quality evaluation of ultrasound imaging in the carotid artery based on normalization and speckle reduction filtering. Med. Bio. Eng. Comp. **44**, 414 (2006)

12. Saeed, S.U., et al.: Learning image quality assessment by reinforcing task amenable data selection. In: Feragen, A., Sommer, S., Schnabel, J., Nielsen, M. (eds.) IPMI 2021. LNCS, vol. 12729, pp. 755–766. Springer, Cham (2021). https://doi.org/10.1007/978-3-030-78191-0_58

13. Yoon, J., Arik, S., Pfister, T.: Data Valuation using Reinforcement Learning. arXiv: 1909.11671 (2020)

14. Duan, Y., Schulman, J., Chen, X., Bartlett, P.L., Sutskever, I., Abbeel, P.: RL2: Fast Reinforcement Learning via Slow Reinforcement Learning. arXiv: 1611.02779 [cs.AI] (2016)

15. Wang, J.X., et al.: Learning to reinforcement learn. arXiv: 1611.05763 [cs.LG] (2017)

16. Cotter, N.E., Conwell, P.R.: Fixed-weight networks can learn. In: 1990 IJCNN International Joint Conference on Neural Networks, vol. 3, pp. 553–559 (1990)

17. Santoro, A., Bartunov, S., Botvinick, M., Wierstra, D., Lillicrap, T.: Meta-learning with memory-augmented neural networks. In: Balcan, M.F., Weinberger, K.Q. (eds.) Proceedings of The 33rd International Conference on Machine Learning, vol. 48, pp. 1842–1850. Proceedings of Machine Learning Research. PMLR, New York, New York, USA (2016)

18. Younger, A.S., Conwell, P.R., Cotter, N.E.: Fixed-weight on-line learning. IEEE Trans. Neural Networks **10**(2), 272–283 (1999). https://doi.org/10.1109/72.750553

19. Hochreiter, S., Younger, A.S., Conwell, P.R.: Learning to learn using gradient descent. In: Dorffner, G., Bischof, H., Hornik, K. (eds.) ICANN 2001. LNCS, vol. 2130, pp. 87–94. Springer, Heidelberg (2001). https://doi.org/10.1007/3-540-44668-0_13

20. Prokhorov, D.V., Feldkarnp, L.A., Tyukin, I.Y.: Adaptive behavior with fixed weights in RNN: an overview. In: Proceedings of the 2002 International Joint Conference on Neural Networks. IJCNN 2002, vol. 3, pp. 2018–2022 (2002). https://doi.org/10.1109/IJCNN.2002.1007449

21. Botvinick, M., Ritter, S., Wang, J.X., Kurth-Nelson, Z., Blundell, C., Hassabis, D.: Reinforcement learning, fast and slow. Trends Cogn. Sci. **23**(5), 408–422 (2019)

22. Schulman, J., Wolski, F., Dhariwal, P., Radford, A., Klimov, O.: Proximal Policy Optimization Algorithms. arXiv: 1707.06347 [cs.LG] (2017)

23. Nichol, A., Achiam, J., Schulman, J.: On First-Order Meta-Learning Algorithms. arXiv: 1803.02999 [cs.LG] (2018)
24. Krizhevsky, A., Sutskever, I., Hinton, G.: ImageNet classification with deep convolutional neural networks. In: NeurIPS (2012)
25. Ronneberger, O., Fischer, P., Brox, T.: U-Net: convolutional networks for biomedical image segmentation. In: Navab, N., Hornegger, J., Wells, W.M., Frangi, A.F. (eds.) MICCAI 2015. LNCS, vol. 9351, pp. 234–241. Springer, Cham (2015). https://doi.org/10.1007/978-3-319-24574-4_28

Deep Video Networks for Automatic Assessment of Aortic Stenosis in Echocardiography

Tom Ginsberg[1]([⊠]), Ro-ee Tal[1], Michael Tsang[3], Calum Macdonald[1],
Fatemeh Taheri Dezaki[2], John van der Kuur[1], Christina Luong[3],
Purang Abolmaesumi[2], and Teresa Tsang[3]

[1] Engineering Physics Project Lab, University of British Columbia,
Vancouver, Canada
tom.ginsberg@alumni.ubc.ca
[2] Department of Electrical and Computer Engineering, University of British
Columbia, Vancouver, Canada
[3] Vancouver General Hospital Echocardiography Laboratory, Vancouver, Canada

Abstract. Aortic valve stenosis (AS) is the narrowing of the heart's aortic valve opening, which restricts blood flow from the left ventricle to the aorta. Accurate diagnosis and timely intervention of AS are crucial since the mortality rate of this condition rapidly increases as symptoms begin to develop. Automated AS estimation in echocardiography faces several challenges, including generalization from diverse medical data, access to high-quality Doppler imaging, and noisy training labels. In this paper, we propose a method for automatic Aortic Stenosis assessment in echocardiography, which is, to the best of our knowledge, the first deep learning pipeline to automate the identification and grading of AS using cardiac ultrasound. Trained and evaluated on a large dataset of 9,117 echocardiograms obtained from 2,247 patients, our method achieves a mean F_1 score of 96.5% for the identification of AS and a mean F_1 score of 73% for grading AS. We use a multi-task training scheme to predict AS severity and key parameters used in clinical AS assessment along with their aleatoric uncertainties. Compared to a baseline that only predicts AS severity, our results show that our multi-task uncertainty-aware inference method achieves comparable classification performance while improving the ability to detect out-of-distribution examples. This is crucial for the clinical deployment of our method in point-of-care settings, where ultrasound operators have less experience in acquiring high-quality echocardiograms.

Keywords: Deep learning · Echocardiography · Ultrasound · Uncertainty estimation · Multi-task learning · Point-of-care

---

T. Ginsberg, R. Tal and M. Tsang—Joint first authors.
P. Abolmaesumi and T. Tsang—Joint senior authors.

© Springer Nature Switzerland AG 2021
J. A. Noble et al. (Eds.): ASMUS 2021, LNCS 12967, pp. 202–210, 2021.
https://doi.org/10.1007/978-3-030-87583-1_20

1   Introduction

1.1   Clinical Background: Aortic Stenosis

Aortic stenosis (AS) is the most prevalent and deadly valvular heart disease, and is of epidemic proportion in developed countries [3]. The increased prevalence of AS with age poses considerable public health problems for aging populations. A recent study demonstrated that the number of hospitalizations related to AS increased by 43% between 2004 and 2013, particularly among patients aged above 85 years [8]. Although AS is initially characterised by a prolonged asymptomatic period, the mortality rate associated with this disorder rapidly increases as symptoms begin to develop. Moreover, while most patients who undergo aortic valve intervention demonstrate long-term survival similar to the general population [2], only one third of untreated patients remain alive after 5 years [19]. Hence, early diagnosis and timely intervention are crucial for clinical AS management.

The clinical standard for AS assessment involves two-dimensional echocardiographic (echo) imaging of the aortic valve, as well as Doppler imaging for the derivation of key hemodynamic parameters. The parasternal long-axis (PLAX) and parasternal short-axis at the aortic valve level (PSAX-Ao) are the most frequently used two-dimensional echo imaging windows for the subjective assessment of AS severity because the opening and closing of the aortic valve can be visualized in real time. AS severity is typically characterized by three parameters: the calculated *aortic valve area* (AVA), *peak aortic velocity*, and *mean transaortic pressure gradient*. The AVA is calculated based on the fluid continuity equation, which involves the measured diameter of the left ventricular outflow tract (LVOT), the time-velocity integral (TVI) of the LVOT by pulsed-wave Doppler, and the aortic valve TVI by continuous-wave Doppler. Erroneous measurement of these parameters will introduce inaccuracies into the estimation of AVA, and the AS assessment itself. Previous studies have demonstrated that discordant grading of aortic stenosis (Table 1) can be observed in up to 30% of patients whose calculated AVA does not match the AS severity suggested by the mean transaortic gradient [6]. Strategies for the grading AS severity without relying on additional parameters provided by Doppler imaging may help improve the accuracy and precision of grading of AS severity in specialized echo laboratories.

1.2   Our Contributions

In this paper, we propose the application of state-of-the-art deep learning techniques with a large-scale dataset to predict aortic stenosis from echo, without the need for specialized Doppler imaging for inference. We attempt to leverage task-specific aleatoric uncertainties in a multi-task paradigm to address the pitfalls of standard clinical grading. In summary, our key contributions are:

1. To the best of our knowledge, the first deep learning implementation for automatic grading of aortic stenosis in echocardiography;

2. A multi-task, uncertainty-aware training scheme well suited for medical data with potentially absent auxiliary labels;
3. An evaluation of the strengths and limitations of our method to guide further research in automated AS classification in the point-of-care setting.

1.3 Related Work

Video Classification: It is natural to apply established deep learning architectures for video directly to echo data, since echo is conventionally captured as a sequence of images. A simple but effective method for video classification is to replace conventional 2D convolutions in CNN architectures with 3D spatio-temporal or notably superior 2D spatial followed by 1D temporal (2+1D) convolutions [20]. The most recent architectures achieving state-of-the-art performance on benchmarks, such as the Kinetics 400 action recognition dataset, include non-local neural networks [22] and spatio-temporal self-attention [1]. Several recent works have applied video architectures for automatic assessment and measurement in echo data. Most notable is Echo-Net Dynamic [17], which shows the effectiveness of an 18 layer 2+1D convolutional residual network for the estimation of left ventricular ejection fraction (LVEF)—a common indicator of cardiac function. Additional works include segmentation of cardiac chambers [5], synchronization of views [9], detection of landmarks [13], and image view and quality estimation [15].

Multi-task Learning and Uncertainty Assessment: Multi-task learning is an approach that aims to improve generalization by using the information, or shared representations, derived from learning related objectives as inductive biases [4]. This is usually conducted via hard parameter sharing (a subset of parameters are shared between all tasks while others are task specific), but can also be accomplished with soft parameter sharing (all parameters are task specific but are jointly constrained by priors) [18]. Multi-task learning has proven successful across many domains, including computer vision [24], and offers several advantages including data efficiency, reduced over-fitting and faster learning [7].

Automatic and robust uncertainty estimation on model predictions has become ubiquitous in modern machine learning, especially within medical applications. Aleatoric uncertainty (uncertainty in the data) is conventionally modelled by placing a Gaussian prior over a model's outputs and training with a maximum likelihood objective [14]. Additional work has explored estimating aleatoric uncertainty using test-time augmentations [21]. Epistemic uncertainty (uncertainty in the model parameters) can be estimated using deep ensembles [11], Monte Carlo dropout [12] or Bayesian Neural Networks which place prior distributions directly over a model's weights [16].

2   Materials and Method

2.1   AS Dataset

The data used for the experiments was collected from the Picture Archiving and Communication System at Vancouver General Hospital (VGH), Canada, following approval from the Medical Research Ethics Board in coordination with the privacy office. Our dataset contains 9,117 echos gathered from 2,247 patients between 2009 and 2015. Labels for this dataset include the three Doppler measurements described in Table 1 as well as supplementary details on the presence of aortic valve defects, cardiac rhythm, study date, video frame-rate, average heart-rate and the machine type used for acquisition. A subset balanced between all severity classes is selected for training and evaluation containing 4,548 PLAX and 3,576 PSAX-Ao echos obtained from 2,092 patients which includes all normal studies (noted by the clinician as having an unrestricted tricuspid aortic valve) and those with concordant AS labels as defined by Table 1. To alleviate negative biases during training and evaluation all studies with discordant AS labels are excluded from our dataset. The remaining subset containing 993 echos with known aortic valve abnormalities is used as a test set for evaluating model robustness, which we describe further in Subsect. 3.1.

Table 1. The Aortic Valve Area (AVA), Peak Aortic Velocity (PV), and Mean Transaortic Pressure Gradient (MG) are the three primary measurements used for grading AS. A grading is considered concordant if the AVA and one of the PV or MG fall into the same severity level. Discordant grading, which often results from measurement error, is a significant problem in the conventional assessment of AS.

Severity	AVA (cm^2)	PV (m/s)	MG (mmHg)
Mild	>1.5	≤3	≤20
Moderate	1–1.5	3–4	20–40
Severe	≤1	>4	>40

2.2   Modelling Prediction Uncertainty with Gaussian Priors

A common technique for enhancing model interpretability is to incorporate uncertainty in predictions. This is usually accomplished by predicting a distribution over the output space. For regression tasks, this is accomplished by maximum likelihood estimation using a Gaussian likelihood function [14], leading to the negative log-likelihood (NLL) loss function:

$$\mathcal{L}_{\text{NLL}}(\hat{\mu}, \hat{\sigma}, y) = -\log(L(\hat{\mu}, \hat{\sigma} \,|\, y)) = \frac{1}{2\hat{\sigma}^2}(y - \hat{\mu})^2 + \frac{1}{2}\log(2\pi\hat{\sigma}^2). \quad (1)$$

Here, $\hat{\mu}$, $\hat{\sigma}$ are the model outputs defining the mean and standard deviation of the Gaussian likelihood function L.

For the task of AS classification, we cannot make use of this regression loss directly. However, the ordered nature of our classes leads us to desire a loss that preserves their natural distance relative to each other. Hence, we turn to the method of a cumulative Gaussian (CDF) loss [15], which aims to maximize the probability mass of the predicted normal distribution over a target region $[y_1, y_2]$ of the output space:

$$\mathcal{L}_{\text{CDF}}\left(\hat{\mu}, \hat{\sigma}, y_1, y_2\right) = -\log\left(\mathbb{P}(y_1 \leq \hat{y} \leq y_2)_{\hat{y} \sim \mathcal{N}(\hat{\mu}, \hat{\sigma})}\right)$$

$$= -\log\left(\frac{1}{2}\text{erf}\left(\frac{\hat{\mu} - y_1}{\sqrt{2}\hat{\sigma}}\right) - \frac{1}{2}\text{erf}\left(\frac{\hat{\mu} - y_2}{\sqrt{2}\hat{\sigma}}\right)\right). \quad (2)$$

Here, we create a contiguous set of regions (illustrated on the left in Fig. 1), each corresponding to a prediction class.

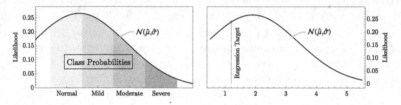

Fig. 1. Comparison of cumulative Gaussian prediction (left) with Gaussian maximum likelihood (right). Note that the model output spaces are identical. In the cumulative case, the probability mass in each region corresponds to the predicted class probability.

2.3   Implementation

Neural Network Architecture: The backbone of our model is an 18 layer deep residual network (ResNet18) with spatio-temporal convolutions. This architecture has been validated by previous work as the gold-standard for video classification and regression tasks with echo [10,17]. In the multi-task setting, this backbone comprises the shared parameters of the network. The features extracted from the backbone are passed to individual fully connected layers which predict outputs for each auxiliary training task.

Training: We follow standard multi-task learning (MTL) methods and train using stochastic gradient-based optimization methods over the unweighted sum of losses for all task predictions to jointly learn both the shared and task-specific parameters (hard parameter sharing) [23]. While newer MTL training methods have been proposed [25], the focus of this paper is to use MTL in a novel application for automated ultrasound analysis and interpretation. The input to our model is approximately one cardiac cycle from an echo that is then resampled using trilinear interpolation to 16 frames and a spatial dimension of 224×224. We use Adam optimization with a base learning rate of 10^{-4} and a learning rate decay factor of 0.1 every 15 epochs. Using a batch size of 8 on a single Nvidia TITAN RTX GPU, we train, validate and test our model on 78%, 12%, and 10% random splits of our dataset, respectively. We train for a maximum of 60 epochs, validating at the end of each epoch. We perform random square cropping and horizontal flipping with a probability of 0.3. The best model is selected by validation performance. Echo data is preprocessed by zero masking the region outside of the ultrasound beam area. During training, our model dynamically handles missing auxiliary labels by zeroing the loss of the relevant tasks, which is necessary for incompletely-labelled medical data.

3   Experiments and Results

We begin with several ablation studies, aiming to validate the efficacy of a) AS identification and grading, b) modelling aleatoric uncertainty with Gaussian predictions for AS echo data, and c) multi-task learning for this application. We summarize our main results in Table 2. By selecting the most commonly predicted class for echos belonging to the same patient study (a strategy known as majority voting, or MV), we demonstrate improved inference ability. This performance boost overcomes the slight performance decrease we observe when training on both PLAX and PSAX data (bottom two rows). Figure 2 illustrates the multi-task model class predictions for both AS grading and identification. For the task of AS identification, where we group mild, moderate and severe AS into a single class, we achieve a mean F_1 score of 96.3% on our multi-task model trained using Gaussian objectives (see Subsect. 2.2). For the task of severe AS identification, where we group normal, mild AS and moderate AS into a single class, we achieve a mean F_1 score of 84.6% on the same model. Our results also confirm prior research [14] that claims improved training robustness on noisy data and performance increases associated with uncertainty predictions. While we observe little negative transfer of learning in the multi-task setting, predicting clinically relevant auxiliary labels can support clinical decision making, such as confirming the AVA falls within an expected range (see Fig. 3).

Table 2. Evaluation benchmarks on aortic stenosis (AS) grading. We demonstrate improved performance using Gaussian objectives and show that using majority voting (MV) on predictions pertaining to the same patient study improves inference performance, especially when trained on both PLAX and PSAX data (bottom two rows).

Learning task(s)	Loss	Recall (%)		F_1 (%)		Recall (%) MV	F_1 (%) MV	F_1 (%) AS ID[a]
		PLAX	PSAX	PLAX	PSAX			
AS (PLAX)	Cross-Entropy	61.5	–	61.3	–	61.8	61.9	90.4
AS (PLAX)	Gaussian	67.9	–	67.9	–	69.6	69.7	97.5
AS	Gaussian	66.5	68.3	66.8	68.8	73.0	73.2	96.8
AS+AVA	Gaussian	66.9	66.9	67.0	67.2	71.0	71.1	96.3

[a] Here we compute the F_1 score for binary AS identification with majority voting.

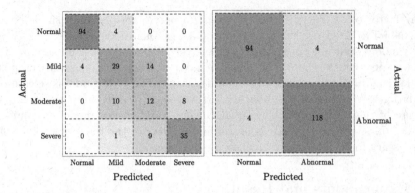

Fig. 2. Confusion matrices for the majority voting inference strategy applied to our multi-task model trained on PSAX/PLAX with Gaussian loss functions. The evaluation is performed on 220 studies corresponding to 10% of our total dataset. We note that the confusion of our model between mild and moderate AS presents similarly to real clinical practice.

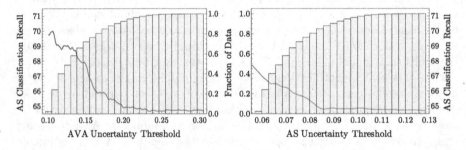

Fig. 3. Aortic stenosis (AS) classification performance vs. prediction uncertainty threshold for two strategies. Each plot shows the average recall on AS classification for predictions where the task uncertainty falls below a given threshold. In the background of each plot is a histogram showing the fraction of total predictions kept for a particular threshold. Note thresholding based on the aleatoric uncertainty for the aortic valve area (AVA) task improves performance at a faster rate compared to thresholding based on AS uncertainty.

3.1   Out-of-Distribution Detection

Recognizing the inherent challenge in generalizing from diverse echo data (which is reflected in our results), we feel it is necessary to highlight some of the unique features of our data and showcase our performance in identifying out-of-distribution samples. Most notably, Aortic valves may possess a varying number of leaflets, which leads to fundamental differences in their shape and structure, and demands specialized AS assessment. Training solely on cases with tricuspid (regular) aortic valves, we demonstrate considerable out-of-distribution detection when predicting aleatoric uncertainties on bicuspid cases, which we visualize in

Fig. 4. We find similar results analyzing the uncertainty predictions of the AVA regression task.

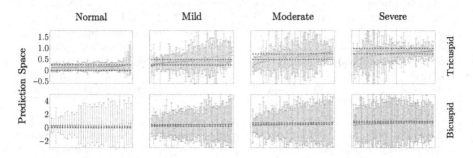

Fig. 4. Aortic stenosis evaluation predictions for tricuspid (top row) and bicuspid (bottom row) cases. The target prediction region for each class is displayed as a light green band enclosed between dashed lines. Vertical interval markers show the model prediction and uncertainty for each study. Bicuspid aortic valves are uncommon and represent one type of out-of-distribution example. Significantly higher uncertainty is predicted for the bicuspid cases. (Color figure online)

4   Conclusion

Timely and accurate AS assessment and intervention are becoming increasingly vital to healthcare systems in the developed world. Our findings showcase the efficacy of modern deep learning methods for automatic AS identification and grading using only cardiac ultrasound. Moreover we demonstrate a multi-task, uncertainty-aware inference methodology that provides clinically relevant predictions by identifying and excluding out-of-distribution examples. Our method also presents an initial step towards democratizing AS assessment, as it does not rely on specialized echocardiography at inference time, supports multiple echo views, and has the potential to be deployed to portable ultrasound devices. While we evaluate our methodology only on AS data, our proposed strategy is generic and applicable to other medical imaging domains where multiple data labels are often present and potentially missing.

References

1. Bertasius, G., Wang, H., Torresani, L.: Is space-time attention all you need for video understanding? (2021)
2. Brown, M.L., et al.: The benefits of early valve replacement in asymptomatic patients with severe aortic stenosis. J. Thorac. Cardiovasc. Surg. **135**(2), 308–315 (2008)
3. Carabello, B.A., Paulus, W.J.: Aortic stenosis. Lancet **373**(9667), 956–966 (2009)
4. Caruana, R.: Multitask learning. Learning to Learn, pp. 95–133 (1998)

5. Chen, C., et al.: Deep learning for cardiac image segmentation: a review. Front. Cardiovasc. Med. **7**, 25 (2020)

6. Clavel, M.A., et al.: The complex nature of discordant severe calcified aortic valve disease grading: new insights from combined doppler echocardiographic and computed tomographic study. J. Am. Coll. Cardiol. **62**(24), 2329–2338 (2013)

7. Crawshaw, M.: Multi-task learning with deep neural networks: a survey (2020)

8. Czarnecki, A., et al.: Trends in the incidence and outcomes of patients with aortic stenosis hospitalization. Am. Heart J. **199**, 144–149 (2018)

9. Dezaki, F.T., et al.: Echo-SyncNet: self-supervised cardiac view synchronization in echocardiography. IEEE Trans. Med. Imag. **40**, 2092–2104 (2021). https://doi.org/10.1109/TMI.2021.3071951

10. Kazemi Esfeh, M.M., Luong, C., Behnami, D., Tsang, T., Abolmaesumi, P.: A deep Bayesian video analysis framework: towards a more robust estimation of ejection fraction. In: Martel, A.L., et al. (eds.) MICCAI 2020. LNCS, vol. 12262, pp. 582–590. Springer, Cham (2020). https://doi.org/10.1007/978-3-030-59713-9_56

11. Fort, S., Hu, H., Lakshminarayanan, B.: Deep ensembles: a loss landscape perspective (2020)

12. Gal, Y., Ghahramani, Z.: Dropout as a Bayesian approximation: representing model uncertainty in deep learning (2016)

13. Jafari, M.H., et al.: U-LanD: uncertainty-driven video landmark detection (2021)

14. Kendall, A., Gal, Y.: What uncertainties do we need in Bayesian deep learning for computer vision? CoRR abs/1703.04977 (2017)

15. Liao, Z., et al.: On modelling label uncertainty in deep neural networks: automatic estimation of intra-observer variability in 2d echocardiography quality assessment (2019)

16. Mitros, J., Namee, B.M.: On the validity of Bayesian neural networks for uncertainty estimation (2019)

17. Ouyang, D., et al.: Video-based AI for beat-to-beat assessment of cardiac function. Nature **580**(7802), 252–256 (2020)

18. Ruder, S.: An overview of multi-task learning in deep neural networks (2017)

19. Strange, G., et al.: Poor long-term survival in patients with moderate aortic stenosis. J. Am. Coll. Cardiol. **74**(15), 1851–1863 (2019)

20. Tran, D., Wang, H., Torresani, L., Ray, J., LeCun, Y., Paluri, M.: A closer look at spatiotemporal convolutions for action recognition (2018)

21. Wang, G., Li, W., Aertsen, M., Deprest, J., Ourselin, S., Vercauteren, T.: Aleatoric uncertainty estimation with test-time augmentation for medical image segmentation with convolutional neural networks. Neurocomputing **338**, 34–45 (2019)

22. Wang, X., Girshick, R., Gupta, A., He, K.: Non-local neural networks (2018)

23. Wu, S., Zhang, H.R., Ré, C.: Understanding and improving information transfer in multi-task learning. arXiv e-prints arXiv:2005.00944, May 2020

24. Zamir, A., Sax, A., Shen, W., Guibas, L., Malik, J., Savarese, S.: Taskonomy: disentangling task transfer learning (2018)

25. Zamzmi, G., Rajaraman, S., Antani, S.: UMS-Rep: unified modality-specific representation for efficient medical image analysis. Inform. Med. Unlocked **24**, 100571 (2021). https://doi.org/10.1016/j.imu.2021.100571. https://www.sciencedirect.com/science/article/pii/S2352914821000617

Pruning MobileNetV2 for Efficient Implementation of Minimum Variance Beamforming

Sobhan Goudarzi[1]($\boxtimes$) (ID), Amir Asif[2] (ID), and Hassan Rivaz[1] (ID)

[1] Department of Electrical and Computer Engineering, Concordia University, Montreal, QC, Canada
sobhan.goudarzi@concordia.ca, hrivaz@ece.concordia.ca
[2] Department of Electrical Engineering and Computer Science, York University, Toronto, ON, Canada
asif@eecs.yorku.ca

Abstract. Beamforming is an essential step in ultrasound image reconstruction that can alter both image quality and framerate. Adaptive methods estimate a set of data-dependent apodization weights among which Minimum Variance Beamforming (MVB) is one of the most powerful approaches performing well regardless of the imaging settings. MVB, however, is not applicable online as it is computationally expensive. Recently, in order to speed up MVB, we took advantages of state-of-the-art methods in deep learning and adapted the MobileNetV2 structure to train and test a model that mimics MVB. In terms of image quality, our method was ranked first in the Challenge on Ultrasound Beamforming with Deep Learning (CUBDL). However, considering both image quality and network size, our method was jointly ranked first with another submission which had a smaller number of parameters. The number of parameters and processing time are important especially for the point-of-care ultrasound machines, which have limited size and computational power. Herein, we propose an approach to prune the trained MobileNetV2 to reduce the number of parameters and computational complexity to further speed-up beamforming. Results confirm that there is no discernible reduction in the network performance, in terms of either visual or quantitative comparison after pruning. In terms of the memory footprint, the post-pruned networks contain 0.3 million parameters as compared to the 2.3 million pre-pruned networks, a reduction by a factor of 7.67. The run-times of MVB, pre- and post-pruned models are 4.05, 0.67 and 0.29 min, respectively.

Keywords: Neural network pruning · Adaptive beamforming · Ultrasound imaging · Minimum variance · Deep learning

1 Introduction

Medical ultrasound image formation pipeline is designed to reconstruct a high-quality spatial map of the target echogenicity. Receive beamforming is one of

J. A. Noble et al. (Eds.): ASMUS 2021, LNCS 12967, pp. 211–219, 2021.
https://doi.org/10.1007/978-3-030-87583-1_21

the essential steps which refers to the process of combining the outputs of different crystal elements of the probe. The goal is to estimate the pixel intensities corresponding to averaged tissue reflectivity functions over the range of medium under consideration. Delay and Sum (DAS) is the most common approach in which, first, the propagation delays corresponding to each signal recorded by piezoelectric elements are compensated. Afterward, a set of element weights, referred to as an apodization, are used for information fusion among different crystal elements through weighted summation.

In contrast to DAS, in which predetermined apodization weights are used, adaptive approaches directly estimate the weights from channel data in order to improve the resolution and contrast. Minimum Variance Beamforming (MVB) is one of the well-known adaptive methods which provides excellent image quality regardless of the imaging settings. MVB estimates the apodization weights minimizing the output variance while preserving the unity gain in the steering direction [15]. Nevertheless, MVB is not applicable online as it requires estimation of the covariance matrix of data which is very time-consuming. Therefore, a growing body of research has addressed speeding-up MVB to make it applicable real-time [1,4,6,16].

During the past few years, deep learning has been utilized for the low-level task of ultrasound image reconstruction [5,9,10]. Recently, the Challenge on Ultrasound Beamforming with Deep Learning (CUBDL) was held in conjunction with the 2020 IEEE International Ultrasonics Symposium (IUS) [2,8]. We successfully participated in this challenge and proposed a general approach for ultrasound beamforming using deep learning [6]. More specifically, the MobileNetV2 [14] structure was adapted to train a model that mimics Minimum Variance Beamforming (MVB). In terms of image quality, our method was ranked first. Overall, considering both image quality and network size, our method was jointly ranked first with another submission [12].

Unfortunately, deep models require a large memory and are computationally expensive, which not only increase the hardware costs, but also precludes them from practical application on resource-constrained environments such as the point-of-care ultrasound machines. Neural network pruning is a common approach for solving these issues. Pruning involves the systematic removing of network parameters to produce a smaller model with similar performance [3]. Herein, we propose an approach to prune the trained MobileNetV2 to reduce the number of parameters and computational complexity to speed-up beamforming.

In this study, the L1 norm is used to score the importance of each parameter for pruning. The rationale behind this selection is that kernels with a smaller norm have relatively lower impact on the network's output. Starting from the last layer (i.e., the layer closest to the network output), the pruning processes backwards towards the first layer. More specifically, pruning starts with removing a specific fraction (α) of the kernels with the lowest $L1$ norms in the last layer. The same procedure is then successively applied to the earlier layers until the first layer is reached. While pruning each layer, we start with a small value of α and gradually increase it until a reduction in the output's quality is observed.

We upper bound α to the maximum value of 0.7. Experimental results confirm that there is no discernible reduction in the network performance, in terms of either visual or quantitative comparison, after pruning. In terms of the memory footprint, the pre- and post-pruned networks contain 2.3 and 0.3 million parameters (7.67 times smaller with pruning). The run-times of MVB, pre- and post-pruned models are 4.05, 0.67 and 0.29 min, respectively. The contributions of this paper are summarized below:

1. To the best of our knowledge, we exploit pruning for the first time in ultrasound beamforming, and more generally in ultrasound imaging.
2. We propose to utilize the L1 norm to prune our network by first pruning from deep layers and then gradually increasing the pruning factor α.
3. We propose the first MVB-based deep beamformer that is approximately 14 times faster than MVB, paving the way for wider use of adaptive beamforming in real-time ultrasound applications.

The rest of the paper is organized as follows. First, the beamforming model based on MobileNetV2 is explained in Sect. 2. The proposed pruning algorithm is introduced in Sect. 3. Results are presented and discussed in Sect. 4. Finally, the paper is concluded in Sect. 5.

2   Beamforming Using MobileNetV2

Let us assume an ultrasound array that transmits a pulse into the medium with a sound speed of c. Without loss of generality, consider n elements record the backscattered signals denoted by $\mathbf{h}_i(t)$. The transmission distance between the origin of the transmitted pulse to an arbitrary pixel (x, z) in the region-of-interest (ROI) is denoted by d_t. Likewise, d_r is defined as the receiving distance from (x, z) to the location of element i. The radio-frequency (RF) data corresponding to (x, z) in $\mathbf{h}_i(t)$ can be determined using the propagation delay as follows (hereafter, capital and bold font variables represent matrices and vectors, respectively):

$$\tau(x, z) = \frac{d_t + d_r}{c} \implies S_i(x, z) = \mathbf{h}_i(t) \,|_{t=\tau(x,z)}, \tag{1}$$

where matrix S_i contains the RF data recorded by crystal element i corresponding to ROI pixels. The final RF image is obtained using a weighted summation of all receiving elements as follows:

$$S(x, z) = \sum_{i=0}^{n-1} \mathbf{w}_i(x, z) S_i(x, z), \tag{2}$$

where $\mathbf{w}$ refers to the apodization window of length n. In practice, the number of active crystal elements considered for the reconstruction of each depth is determined using F-number. Moreover, S is subject to envelope detection and log compression for obtaining the final B-Mode ultrasound image.

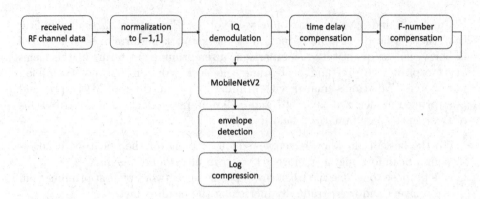

Fig. 1. Diagram of beamforming using MobileNetV2 [6].

In Capon's MVB, $\mathbf{w}$ is adaptively estimated such that the output variance is minimized while the unity gain is preserved in the steering direction [15]. The final solution of MVB is as follows [15]:

$$\mathbf{w}_{MV} = \frac{R^{-1}\mathbf{a}}{\mathbf{a}^H R^{-1}\mathbf{a}} \tag{3}$$

where R is the spatial covariance matrix. For delayed signals, the steering vector $\mathbf{a} = 1$. The robustness in estimating R is increased with temporal averaging over $2k+1$ samples followed by another averaging over subarrays of length l as follows [15]:

$$\widetilde{R}(x,z) = \frac{\sum_{j=-k}^{k}\sum_{i=0}^{n-l}\bar{\mathbf{r}}_i(x,z-j)\bar{\mathbf{r}}_i^H(x,z-j)}{(2k+1)(n-l+1)} \tag{4}$$

where:

$$\bar{\mathbf{r}}_i(x,z) = \left[S_i(x,z), S_{i+1}(x,z), ..., S_{i+l-1}(x,z)\right]^T \tag{5}$$

A diagonal loading factor is added to the covariance matrix for numerical stability by $\widehat{R}(x,z) = \widetilde{R}(x,z) + \epsilon I$, where I is the identity matrix and:

$$\epsilon = \frac{\Delta}{l}trace(\widetilde{R}(x,z)) \tag{6}$$

The result of subarray averaging is a vector of length l. Finally, each pixel (x,z) of S, using MVB, can be computed as follows:

$$S_{MV}(x,z) = \frac{1}{n-l+1}\sum_{i=0}^{n-l}\mathbf{w}_{MV}^H\bar{\mathbf{r}}_i, \tag{7}$$

Our deep learning framework for ultrasound beamforming [6] based on MobileNetV2 structure is summarized in Fig. 1. Similar to MVB, each pixel of the image is reconstructed separately. More specifically, first, the input data

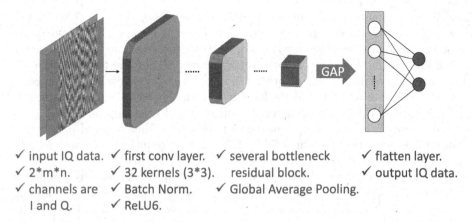

✓ input IQ data.   ✓ first conv layer.   ✓ several bottleneck   ✓ flatten layer.
✓ 2*m*n.   ✓ 32 kernels (3*3).   residual block.   ✓ output IQ data.
✓ channels are   ✓ Batch Norm.   ✓ Global Average Pooling.
  I and Q.   ✓ ReLU6.

Fig. 2. Overview of the beamforming model based on the MobileNetv2 architecture. GAP refers to Global Average Pooling [6].

is scaled to be within the $[-1, 1]$ range. The second step is the IQ demodulation of RF channel data because MVB works on complex signals to estimate complex weights allowing for beampatterns that are asymmetrical around the center of the beam. In the third step, time delay compensation is performed to simplify the mapping for the network. Then, to have a uniform image quality for all depths, F-number is fixed. The network's input is a $2 \times m \times n$ tensor in which n and m are the number of channels and the window length used for temporal averaging to preserve the speckle statistics, respectively. The two channels contain the real and imaginary parts of IQ data. The network is trained to estimate $\mathbf{w}_{MV}$ and apply Eq. (7) to obtain the beamformed IQ data. More details regarding this approach can be found in [6].

3   The Proposed Pruning Method

Herein, a Convolutional Neural Network (CNN) architecture is defined as a function family f, and the trained model is defined as a particular parametrization of f, i.e., $f(x, \theta)$ for specific parameters θ. CNN pruning is the process of taking as input a model $f(x, \theta)$ and generating a smaller model $f(x, \hat{\theta})$ in which $\hat{\theta}$ is a subset of θ.

There are numerous methods of pruning among which the method proposed by Han *et al.* [7] is the most popular one. In this general framework, after training a model, each network's parameter is issued a score, and the pruning is completed based on these scores. As pruning reduces the model performance, the resulting model might be subject to further fine-tuning. The explained pruning and fine-tuning process is often iterated to gradually reduce the network size.

In this study, the network parameters, which are kernel weights of convolutional layers, are scored based on their $L1$ norm. We believe that the impact of a kernel on the network's output is proportional to its norm. It is also observed, in our experiments, that reducing the size of convolutional layers closer to the network output causes relatively smaller changes as compared to making the same reduction in initial layers. Therefore, the proposed pruning approach starts from the convolutional layer closest to the network output and works backwards towards the first layer. Our algorithm is implemented iteratively. More specifically, a specific fraction (α) of the kernels with the lowest $L1$ norm is pruned in each iteration. We start with a small value of α and gradually increase it until a reduction in the model's performance is observed. The value of α is upper bounded to a maximum value of 0.7. The same procedure is then successively applied to the earlier layers until the first layer is reached.

As mentioned in Sect. 1, the goal is to speed-up the previously published beamforming approach. An overview of the beamforming model [6], which is based on the MobileNetv2 architecture [13], is shown in Fig. 2. To stop the iterative pruning process in each layer, specialized ultrasound assessment indexes including Full Width at Half Maximum (FWHM) for resolution measurement and generalized Contrast to Noise Ratio (gCNR) as well as Contrast Ratio (CR) for contrast are calculated.

4   Results and Discussions

In order to evaluate the performance of the proposed pruning method, results on the simulation and experimental phantom datasets which are publicly available through UltraSound ToolBox (USTB) [11] are presented. More specifically, both datasets contain one image of point targets for measuring the spatial resolution, and one image of anechoic cysts for measuring the contrast. Figure 3 provides a visual comparison of the original and pruned beamforming networks. As this figure illustrates, there is no discernible reduction in the network performance after pruning, and both pre- and post-pruned models are able to reconstruct images with the same perceivable quality to that of MVB. The quantitative

Table 1. Quantitative results on simulation and experimental phantom datasets in terms of resolution and contrast indexes. SR and SC refer to simulation resolution and contrast datasets, respectively. ER and EC refer to experimental phantom resolution and contrast datasets, respectively. Subscripts $._A$ and $._L$ refer to axial and lateral directions, respectively.

Dataset	SR		SC		ER		EC	
Index	$FWHM_A$	$FWHM_L$	CR	gCNR	$FWHM_A$	$FWHM_L$	CR	gCNR
DAS	0.4	0.82	−15.15	0.74	0.57	0.88	−13.79	0.57
CPWC	0.39	0.56	−31.44	0.97	0.56	0.56	−25.29	0.87
MV	0.41	0.1	−21.15	0.82	0.59	0.43	−16.74	0.69
MobileNetV2	0.42	0.273	−17.15	0.661	0533	0.773	−15.53	0.55
Pruned network	0.42	0.274	−16.86	0.659	0.531	0.767	−14.59	0.52

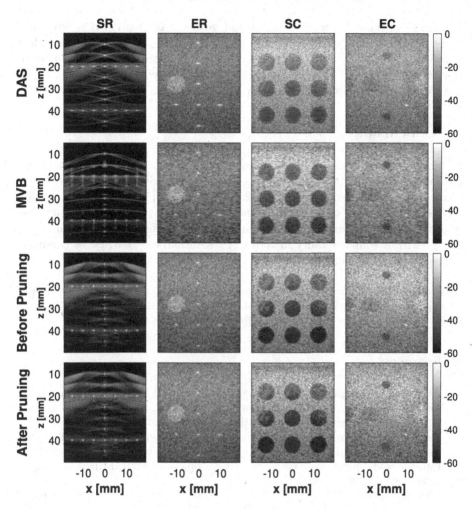

Fig. 3. Beamforming results on the single 0° plane-wave. Columns indicate different image datasets while rows correspond to the beamforming approaches. SR and SC refer to simulation resolution and contrast datasets, respectively. ER and EC refer to experimental phantom resolution and contrast datasets, respectively.

comparison, provided in Table 1, also confirms that there are negligible differences in the values of the calculated indexes in terms of both resolution and contrast between two deep models. In terms of memory footprint, the pre- and post-pruned networks contain 2.3 and 0.3 million parameters, respectively (7.67 times smaller with pruning). For the reconstruction of a single image, the runtimes of MVB, pre- and post-pruned models, are 4.05, 0.67 and 0.29 min, respectively. It should be noted that the deep reconstruction is GPU accelerated. A small memory footprint is of critical importance for commercial success of deep learning beamforming given the very high frame-rate and limited computational resources, especially in mobile ultrasound devices. As mentioned before, similar to MVB, our method reconstructs each pixel of the image separately which makes it quite slow and not close to real time. Therefore, one avenue for future work is to reconstruct the whole image simultaneously to further speed-up the method for real-time implementations.

5   Conclusions

Reduction of network parameters helps speeding up the deep models for beamforming. Herein, the proposed pruning approach reduces the network size by a factor of 7.67 that makes the model 2.3 times faster. As shown in the results, the iterative removal of network weights has been adapted to prevent any loss in performance. The experiments confirm that the pruned network reconstructs images with a similar quality in terms of resolution and contrast as compared to the original beamforming model.

Acknowledgment. We acknowledge the support of the Natural Sciences and Engineering Research Council of Canada (NSERC) RGPIN-2020-04612 and RGPIN-2017-01539. We would like to thank NVIDIA for the donation of the Titan Xp GPU.

References

1. Afrakhteh, S., Behnam, H.: Low-complexity adaptive minimum variance ultrasound beam-former based on diagonalization. Biomed. Sig. Process. Control **62**, 102110 (2020). https://doi.org/10.1016/j.bspc.2020.102110
2. Bell, M.A.L., Huang, J., Hyun, D., Eldar, Y.C., van Sloun, R., Mischi, M.: Challenge on ultrasound beamforming with deep learning (CUBDL). In: 2020 IEEE International Ultrasonics Symposium (IUS), pp. 1–5 (2020). https://doi.org/10.1109/IUS46767.2020.9251434
3. Blalock, D., Ortiz, J.J.G., Frankle, J., Guttag, J.: What is the state of neural network pruning? arXiv preprint arXiv:2003.03033 (2020)
4. Chen, J., Chen, J., Zhuang, R., Min, H.: Multi-operator minimum variance adaptive beamforming algorithms accelerated with GPU. IEEE Trans. Med. Imag. **39**(9), 2941–2953 (2020). https://doi.org/10.1109/TMI.2020.2982239
5. Goudarzi, S., Asif, A., Rivaz, H.: Fast multi-focus ultrasound image recovery using generative adversarial networks. IEEE Trans. Comput. Imag. **6**, 1272–1284 (2020). https://doi.org/10.1109/TCI.2020.3019137
6. Goudarzi, S., Asif, A., Rivaz, H.: Ultrasound beamforming using mobilenetv2. In: 2020 IEEE International Ultrasonics Symposium (IUS), pp. 1–4 (2020). https://doi.org/10.1109/IUS46767.2020.9251565
7. Han, S., Pool, J., Tran, J., Dally, W.: Learning both weights and connections for efficient neural network. In: Advances in Neural Information Processing Systems, vol. 28. Curran Associates, Inc. (2015)
8. Hyun, D., et al.: Deep learning for ultrasound image formation: CUBDL evaluation framework amp;amp; open datasets. IEEE Trans. Ultrasonics Ferroelectrics Frequency Control 1 (2021). https://doi.org/10.1109/TUFFC.2021.3094849

9. Luchies, A.C., Byram, B.C.: Assessing the robustness of frequency-domain ultrasound beamforming using deep neural networks. IEEE Trans. Ultrasonics Ferroelectrics Frequency Control **67**(11), 2321–2335 (2020). https://doi.org/10.1109/TUFFC.2020.3002256

10. Luijten, B., et al.: Adaptive ultrasound beamforming using deep learning. IEEE Trans. Med. Imag. **39**(12), 3967–3978 (2020). https://doi.org/10.1109/TMI.2020.3008537

11. Rodriguez-Molares, A., et al.: The ultrasound toolbox. In: 2017 IEEE International Ultrasonics Symposium (IUS), pp. 1–4 (2017). https://doi.org/10.1109/ULTSYM.2017.8092389

12. Rothlübbers, S., et al.: Improving image quality of single plane wave ultrasound via deep learning based channel compounding. In: 2020 IEEE International Ultrasonics Symposium (IUS), pp. 1–4 (2020). https://doi.org/10.1109/IUS46767.2020.9251322

13. Sandler, M., Howard, A., Zhu, M., Zhmoginov, A., Chen, L.C.: Mobilenetv 2: Inverted residuals and linear bottlenecks. In: Proceedings of the IEEE Conference on Computer Vision and Pattern Recognition, pp. 4510–4520 (2018)

14. Sandler, M., Howard, A., Zhu, M., Zhmoginov, A., Chen, L.C.: Mobilenetv 2: Inverted residuals and linear bottlenecks. In: Proceedings of the IEEE Conference on Computer Vision and Pattern Recognition (CVPR), June 2018

15. Synnevag, J.f., Austeng, A., Holm, S.: Benefits of minimum-variance beamforming in medical ultrasound imaging. IEEE Trans. Ultrasonics Ferroelectrics Frequency Control **56**(9), 1868–1879 (2009). https://doi.org/10.1109/TUFFC.2009.1263

16. Vaidya, A.S., Srinivas, M.: A low-complexity and robust minimum variance beamformer for ultrasound imaging systems using beamspace dominant mode rejection. Ultrasonics **101**, 105979 (2020). https://doi.org/10.1016/j.ultras.2019.105979

Automatic Fetal Gestational Age Estimation from First Trimester Scans

Sevim Cengiz$^{(\boxtimes)}$ and Mohammad Yaqub

Computer Vision, Mohamed bin Zayed University of Artificial Intelligence,
Abu Dhabi, United Arab Emirates
{sevim.cengiz,mohammad.yaqub}@mbzuai.ac.ae
https://mbzuai.ac.ae/biomedia

Abstract. Automatic Gestational Age (GA) estimation based on the Crown Rump Length (CRL) measurement is the preferred solution to overcome the challenges while using the last menstrual period (LMP) to date pregnancies. However, GA estimation based on CRL requires accurate placement of calipers on the fetal crown and rump which is not always a straightforward task, especially for an inexperienced sonographer. This paper proposes an accurate GA estimation method from fetal CRL images during the first trimester scan. The method addresses this problem by segmenting the fetus using a binary and multi-class U-Net. The fetal segmentation is used to compute the CRL. This is then followed by an estimation of GA from the automatic CRL measurement based of clinical information. The results from the multi-class segmentation achieves a more accurate precision, recall, Dice, and Jaccard. This has also led to a more accurate CRL measurement and hence more robust GA estimation.

Keywords: Fetal ultrasound · Deep learning · Crown-rump length · Gestational age estimation · Fetal growth

1 Introduction

All perinatal deaths have been reported with a higher ratio in low-and middle-income countries worldwide [1]. Early detection of gestational development problems might provide indicators of perinatal mortality and morbidity [2,3]. Accurate calculation of fetal gestational age (GA) is important to date the pregnancy, monitor fetal anatomy and growth, and determine the delivery schedule. This is key to detect some possible growth disorders, and preterm birth [4,5].

There are mainly two ways to estimate GA. The first method uses the first day of the woman's last menstrual cycle to date the pregnancy. Due date is estimated after 40 weeks of the last menstrual period (LMP) date. In many cases, LMP is not recorded, or irregular periods make the estimation of GA inaccurate. The second method utilizes ultrasound scanning-based calculation of crown-rump length (CRL) and in some cases measurements on other fetal organs

© Springer Nature Switzerland AG 2021
J. A. Noble et al. (Eds.): ASMUS 2021, LNCS 12967, pp. 220–227, 2021.
https://doi.org/10.1007/978-3-030-87583-1_22

such as the head, femur, and abdomen. Estimation of GA in the first trimester can be done based on the CRL measurement performed on fetuses between 9 and 13 weeks of gestation. Accurate measurement of CRL is paramount for a precise estimation of GA. Although the CRL measurement is in theory relatively straightforward, it is not always accurate especially if performed by a newly qualified sonographer. In addition, the variability in the quality of the acquired ultrasound images makes it challenging to decide the positions for the crown and rump where the calipers need to be placed [6]. Due to the challenges in the LMP estimation method [7], GA estimation based on the CRL measurement between 9–13 weeks is considered a reliable and more accurate option. An accurate GA estimation helps monitor fetal anatomy and growth during the first trimester of pregnancy and it can be used to report the expected date of delivery more precisely. The robustness of GA estimation relies on the accuracy of the CRL measurement. Therefore, reproducible and reliable CRL measurement at early gestation has a good clinical value and may improve patient care.

With the advancements in deep learning methods for medical image assessment, automated computer-based approaches are an increasing trend among researchers. Recently, AI-based calculation of fetal GA was performed using image characteristics and a regression neural network [8]. Another study investigated the identification of fetal imaging planes on prenatal ultrasound, throughout different GAs [9]. Authors in [10] have reported that their machine learning solution may be used to predict preterm birth. The method they proposed segments cervical length (CL) and anterior cervical angle (ACA) and uses these estimations to perform classification for preterm deliveries. A follow-up work, [11] improved the prediction of preterm birth by segmenting and highlighting the cervix using a multi-tasking U-Net network. Another AI-based study was conducted to perform automated brain maturation estimation from 3D ultrasound images by using convolutional regression networks [12].

The main aim of this project is to develop an automatic fetal gestational age estimation method using deep learning. Our method relies on segmenting the fetal head and body accurately to allow for an accurate measurement of the CRL which is then used to estimate fetal GA. Due to the scarcity of public data in this domain, we show extensive evaluation of the developed method and compare the CRL and GA with routine clinical measurements.

2   Materials and Methods

2.1   Dataset

1242 fetal ultrasound images from the first trimester scanning were retrospectively extracted from a hospital archive. All images were from normal pregnancies and anonymized in the hospital before being transferred for further processing. 697 were manually segmented into 2 classes (head and body) by an expert using a drawing annotation tool. Another expert reviewed the manual segmentation and an additional manual review was performed on some images if needed. We call this dataset A, it was used to train the segmentation model and evaluate segmentation accuracy. The remaining 545 images have only CRL measurements performed by a clinician. We call this dataset B and it was used as an independent dataset to assess the accuracy of the CRL measurement and the estimated GA.

2.2   Fetal Head and Body Segmentation

U-Net [13] has proven to work well to segment objects in images in multiple applications. We have investigated multiple U-Net configurations such as image size and network depth. In addition, with cross validation, we investigated different hyper-parameters such as number of epochs, learning rate. The number of epochs, learning rate were 100, 0.00001, respectively. The method was implemented in Python with the TensorFlow library.

The network has a fully convolutional encoder-decoder structure with 4 blocks. Each block is decomposed of multiple convolutional and Relu layers followed by a max pooling layer. We have investigated a binary segmentation U-Net and 3-class (multi-class) segmentation U-Net. In the binary segmentation, a background and a whole fetus classes were used. In the 3-class segmentation U-Net, a background, head, and body classes were used. The U-net was trained by minimizing the sum of the binary cross-entropy loss function and the categorical cross-entropy loss function for binary and multi-class segmentation, respectively.

2.3   Data Augmentation

Due to the small dataset, we performed several augmentation methods during training to ensure more accurate and robust segmentation. We randomly applied horizontal flip, rotation (degrees $[-10, 10]$), and random brightness contrast (α is 20%). Data augmentation was only performed during the training stage. No augmentation was performed during the validation stage.

2.4   Evaluation Metrics

Segmentation Metric. During the training of the segmentation models, 5 fold cross-validation on images from Dataset A was performed. Jaccard, dice, precision, and recall were computed to assess the segmentation performance. Scores were computed for each class background, fetus, head, and body. When reporting the mean accuracy, the background class was not included.

CRL Measurement. A new segmentation model is trained on all images from Dataset A. This model was then applied to images from Dataset B to generate segmentation masks. From the segmentation mask, we calculate the CRL measurement as the longest distance between contour points. First, the contour of the segmentation is computed as the difference between the mask image and a dilated mask with a kernel of size 3×3. Figure 1 shows an example ultrasound image (A), automatic mask with the dilated contour (purple color) (B), and the contour from the automatic and dilated masks (C).

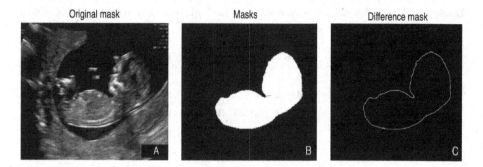

Fig. 1. An example of (A) original ultrasound image (B) automatic mask with the dilated contour (purple color) (B), and the contour from the automatic and dilated masks (C). (Color figure online)

Second, to reduce the computation when finding the longest distance between all contour points, we only consider the contour points in the left and right parts of the fetus. This is based on the assumption that fetuses are typically positioned on their back and hence the CRL is more likely to be a horizontal line. However, this is not always true. In some cases, fetal position is slightly oblique which means that we cannot rely on the furthest two points horizontally. Therefore, we compute the distance between all contour points from the left 20% of the fetus to all contour points from the right 20% of the fetus. Selecting 20% of the contour points on the right and left of the fetus was experimentally chosen, which reduced the computation time by 60%. To optimize this further, some adjacent points could be skipped when finding the distances. The distance between two points is computed as the EUCLIDEAN distance. The maximum euclidean distance is used to measure the CRL in the pixel space. We multiply this distance by the pixel spacing to compute the distance in millimeter. CRL is computed on the segmentation mask in the binary segmentation. However, in the multi-class segmentation, the head and body segmentations are merged to one class which we call "multi-class fetal only" before measuring the CRL.

Gestational Age Estimation. In clinical research, there are multiple published methods which demonstrate how to compute GA from the CRL measurement. We reviewed some of these methods and found that the method developed by [7] was well validated clinically and hence is used in our work. The GA is estimated using Eq. 1,

$$GA = 40.9041 + (3.21585 \times \Delta_{CRL}^{0.5}) + (0.348956 \times \Delta_{CRL})$$ (1)

where Δ_{CRL} is the maximum distance between head and rump in mm.

3   Results

We have investigated many U-Net configurations including different hyperparameter settings. We show that the multi-class segmentation has a better

Table 1. Mean and standard deviation Dice, Jaccard, precision, recall scores of binary and multi-class segmentation.

	Dice	Jaccard	Precision	Recall
Binary classification	93.7 ± 5.6	88.6 ± 8.6	90.5 ± 8.5	97.7 ± 3.8
Multi-class (fetal only)	94.8 ± 6.1	90.7 ± 8.5	96.2 ± 4.6	94.1 ± 8.2
Multi class head	92.3 ± 10.8	87.0 ± 12.9	93 ± 0.09	92.9 ± 12.7
Multi class body	91.5 ± 10.6	85.5 ± 12.6	94 ± 0.08	90.3 ± 12.3

segmentation performance on all metrics compared to binary class segmentation (Table 1). The results show the mean and standard deviation from the 5 fold cross validation on all images in Dataset A. The accuracy of the CRL measurement and estimated GA is reported in Table 2. We show that multi-class segmentation has a lower mean absolute difference between manual and automatic CRL measurement compared to binary segmentation (1.721 ± 4.34 and 3.866 ± 6.79, respectively) in Dataset B (Table 2).

Figure 2 shows visual segmentation results of good (left column), fair (middle column), and failed (right column) segmentation cases of the images, respectively. Ultrasound images and true CRL values provided by the clinicians are shown in Fig. 2(A, D, and G). Binary segmentation masks fused onto ultrasound images and their calculated CRL values are shown in Fig. 2(B, E, H). Multi-class segmentation masks fused onto ultrasound images and calculated CRL values 1 are shown in Fig. 2(C, F, and I).

On average, training time for 100 epochs was done in 20 min on an NVIDIA Quadro RTX 6000 in Ubuntu 18.04.5 LTS. Segmentation time is performed in about 0.5 s.

Figure 3 shows the correlation between the true GA and the estimated GA based on binary (orange color) and multi-class (blue color) methods. The R^2 are 0.71 and 0.86 for binary and multi class methods respectively which shows a better correlation between the manual GA and the multi class-based estimated GA compared to the binary class one.

Table 2. Mean ± standard deviation of absolute difference of the manual and automatic CRL and GA for the binary and multi-class segmentation.

	CRL Diff. in mm	GA Diff. days
Binary classification	3.87 ± 6.79	1.72 ± 4.34
Multi class (fetal only)	2.15 ± 3.73	0.96 ± 2.36

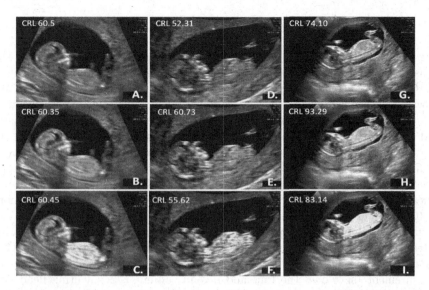

Fig. 2. Examples of good (left column), fair (middle column), and failed (right column) cases of the binary (second row) and multi class (third row) segmentation masks fused onto ultrasound images (first row).

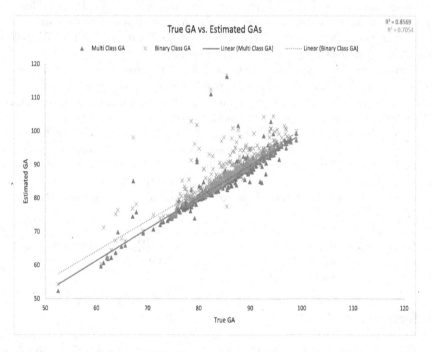

Fig. 3. True GA versus estimated GAs based on binary and multi class segmentation. (Color figure online)

4   Discussion and Conclusion

In this work, we proposed a deep learning segmentation method to segment the fetus in the first trimester scan. We used the segmentation mask to calculate the CRL measurement and subsequently estimate the fetal GA estimation. We showed that segmentation accuracy improves when using multi class instead of binary class U-Net, which also improves the accuracy of the CRL measurement and GA estimation. We believe that the reason behind the better performance of the multi-class segmentation method over binary class is due to the further constraint it imposes during the training stage. In reality, a sonographer performing this task does not only rely on where the fetus is. The sonographer learns what a head and body look like which allow him/her to perform a task like this more accurately. The multi class model we develop follows such an analogy.

Our experiments on Dataset B confirm that the multi-class method performs better than the binary class method for measuring the CRL and estimating GA. Figure 2 shows that the CRL in the multi-class method is close to the human measurement (shown in the first and second columns). Segmenting head and body separately (F) improved the accuracy of measuring the CRL and estimating GA compared to only relying on segmenting the whole fetus (E); this is shown in the second fetal example in Fig. 2 (middle column). On the other hand, for a few cases, both methods failed (right column) to perform well. Imaging features within the head belonging to these few images are in a different position than the normal position, which might have caused both methods to fail. We believe this may be due to the small training set we have.

Figure 3 clearly shows that there are fewer scattered points (outliers) for the multi-class estimation compared to the binary class. In addition, a better correlation with the true GA is shown in Fig. 3 for the multi-class GA estimation ($R^2 = 0.86$) compared to binary class ($R^2 = 0.71$).

One of the limitations of this study is the small dataset size. This may mean that the results we show are not necessarily guaranteed if the method is applied to other datasets. Further evaluation will be needed in the future. In addition, due to the fact that all images we use are from one ultrasound machine type (GE Voluson E8), GA estimation from images acquired on other ultrasound machines requires further assessment.

Although finding the correct imaging plane is clinically important to measure the CRL, this study aimed to only measure the CRL. Future work might be needed to investigate this problem and ensure the correctness of the imaging plane before measuring the CRL.

References

1. Zupan, J.: Perinatal mortality in developing countries. N. Engl. J. Med. **352**, 2047–2048 (2005)
2. Karl, S., et al.: Preterm or not-an evaluation of estimates of gestational age in a cohort of women from Rural Papua New Guinea. PLoS ONE **10**, e0124286 (2015)

3. Rijken, M.J., et al.: Quantifying low birth weight, preterm birth and small-for-gestational-age effects of malaria in pregnancy: a population cohort study. PLoS ONE **9**, e100247 (2014)
4. Alexander, G.R., Tompkins, M.E., Petersen, D.J., Hulsey, T.C., Mor, J.: Discordance between LMP-based and clinically estimated gestational age: implications for research, programs, and policy. Public Health Rep. **110**, 395–402 (1995)
5. Callaghan, W.M., Dietz, P.M.: Differences in birth weight for gestational age distributions according to the measures used to assign gestational age. Am. J. Epidemiol. **171**, 826–836 (2010)
6. Whitworth, M., Bricker, L., Mullan, C.: Ultrasound for fetal assessment in early pregnancy. Cochrane Database Syst. Rev. CD007058 (2015)
7. Papageorghiou, A.T., et al.: International standards for early fetal size and pregnancy dating based on ultrasound measurement of crown-rump length in the first trimester of pregnancy. Fetal International, and Century Newborn Growth Consortium for the 21st. Ultrasound Obstet. Gynecol. **44**, 641–648 (2014)
8. Bradburn, E.H., Hin Lee, L., Noble, J.A., Papageorghiou, A.T.: OC10.04: estimating fetal gestational age based on ultrasound image characteristics using artificial intelligence. Ultrasound Obstetr. Gynecol. **56**, 28–29 (2020)
9. Bradburn, E., Mohammad, Y., Noble, J., Papageorghiou, A.: OC10.05: an artificial intelligence system that can correctly identify fetal ultrasound imaging planes throughout gestational age. Ultrasound Obstet. Gynecol. **56**, 29 (2020). https://doi.org/10.1002/uog.22269
10. Włodarczyk, T., et al.: Estimation of preterm birth markers with U-Net segmentation network. In: Wang, Q., et al. (eds.) PIPPI/SUSI-2019. LNCS, vol. 11798, pp. 95–103. Springer, Cham (2019). https://doi.org/10.1007/978-3-030-32875-7_11
11. Włodarczyk, T., et al.: Spontaneous preterm birth prediction using convolutional neural networks. In: Hu, Y., et al. (eds.) ASMUS/PIPPI -2020. LNCS, vol. 12437, pp. 274–283. Springer, Cham (2020). https://doi.org/10.1007/978-3-030-60334-2_27
12. Namburete, A.I.L., Xie, W., Noble, J.A.: Robust regression of brain maturation from 3D fetal neurosonography using CRNs. In: Cardoso, M.J., et al. (eds.) FIFI/OMIA-2017. LNCS, vol. 10554, pp. 73–80. Springer, Cham (2017). https://doi.org/10.1007/978-3-319-67561-9_8
13. Ronneberger, O., Fischer, P., Brox, T.: U-Net: convolutional networks for biomedical image segmentation. In: Navab, N., Hornegger, J., Wells, W.M., Frangi, A.F. (eds.) MICCAI 2015. LNCS, vol. 9351, pp. 234–241. Springer, Cham (2015). https://doi.org/10.1007/978-3-319-24574-4_28

Author Index

Printed in the United States
by Baker & Taylor Publisher Services